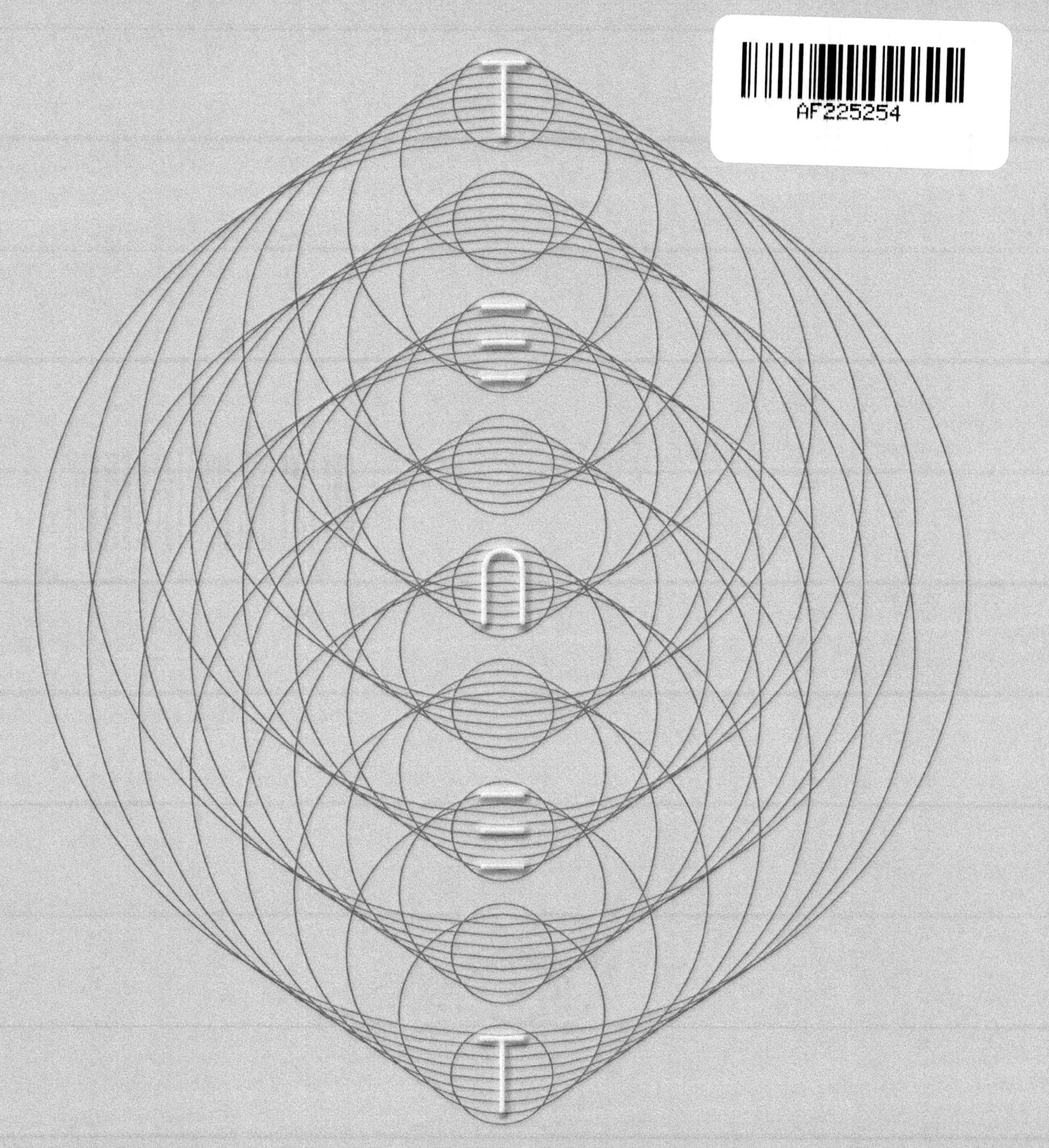
AF225254

38983 : 9862 14 2896

TENET : NEST OF TENS
hiiih   iiiw no wiii
rgngr   ngxo eu ognx
ehehe   eh   r he
et et   t   t

by 10-tentacled antenna

isbn 978-1-940853-52-9

Published by Jalamari Archive in 2025

where 2025 = (1 + 2 + 3 + 4 + 5 + 6 + 7 + 8 + 9)²

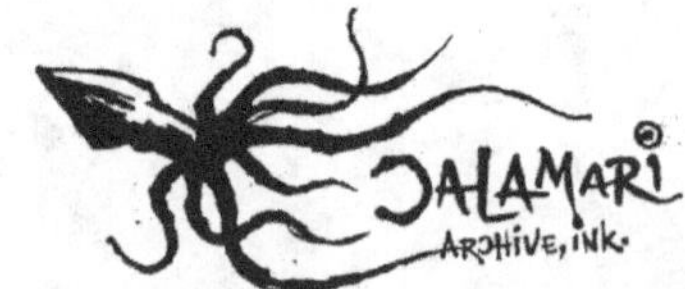

con-10-ts:

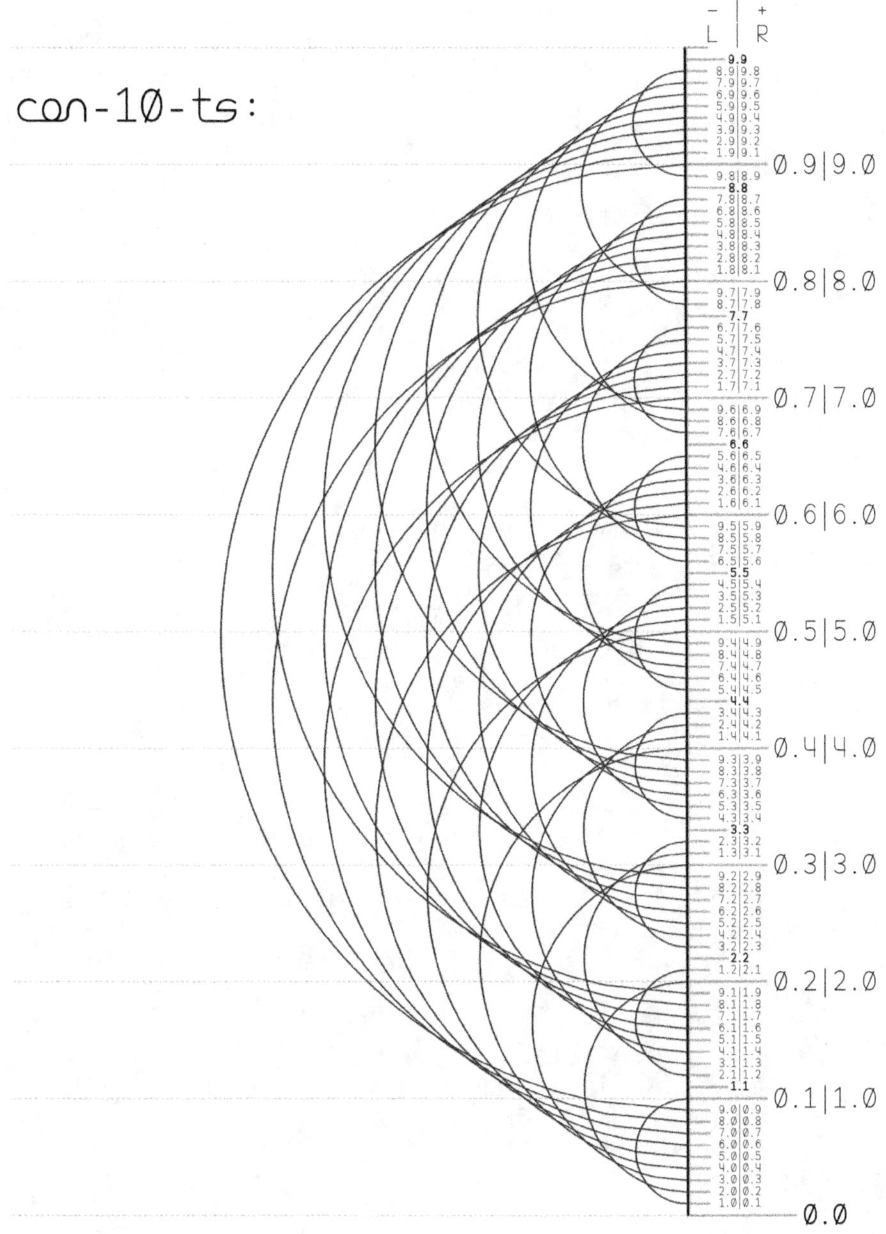

FICTIVE   FICTIVE

T E N E T

Ø . Ø

```
1        [•]        > i.e. 1Ø-ET Ø.Ø of
Ø        (in ^ vaccum)        (mere img of Ø.Ø)        (iRL Ditty)
t    NEST of IDs, 1-off TXT ware (Ø, Ø) = origiNU|| aerea DOOD w-
     here-in i (in8 iD) = 6orn 11/22/66 ; + FT 1Ø-ant teL 1Ø ± 1 yr) + if 1
     types ^ vertical Line: «(Ø, Ø) Lamina tes 8 no dezrever sedit
                    88888888        R        node dotent
t |      -er «sleep Laminates (8 nodes)» Teg U peels» backwoods
         versed if (Ø, Ø)» Starting w/ i-balls, then i (authori-T (2 ^ T)
D +  t   C i = 1     here-in) evolved X bi-fiLogical (rᘓom) dubbed bØd-E
Ø    d Ran   U         type 2 mova said i-6a||s a-
ƚT X   e 6y V = S RROUND X Lᘓscape 2 Zea + [•] fL3E-T 4 (fL3SHED  e  b
P          OoLogical
ƚi  s  e  N    T out) > 4 N Nitre year we R ALL age zerØ 6ut dont   N  e  s
o          (dog 4bird) EggKnow|edge «Ø» az milestØne, fee-
d  1ˢᵗ 8RᗡEB celeB Sum < Set ||UN, Void Spheric UNHoley az «zerØ» Rniŝ
e          K          -REM dont i, caKes c2ted /w 2 or 1 moonᵗʰ
seLf   o   R   ember + mine paRents ≠ alive 2 aSK  eL  impromptU
sYe6aLL   o   even if they still|| = KicKing i wood naught   N
theory m   tRust their taK3 on what happ-   ƚ i
e   6   a   end > reSidual G-Sire reSides iN DNA + «ANd MA i 4
nu||ify   nØ concept of Ø          o          n   M
c   I LiVE 2 3MiT 1/2 LOOPS WOLF H-SELF + TiDES REVERSED          o
D-serve Ø          o          ON 8 SET ANiMAL SLEEP» = 1ˢᵗ Line
D = 5ØØ          m                          1 = i
decimal   in Prior 6ook (iSBN: 978-1-940853-22-2) or Last Line if   N
w   n        o   b        i        T   e        T-ick
e   red 6aKwords + Know Used 2 Seed 2 next? in Keeping 2 mom-   iT
Y 6   i   e   n        6ed   -Entum ({Do geese see goD}) yet @ tH
+   s h   realistic   Same Time Shake old haBiTS + ever-pre-   E
b + W, coPied?          p o          q o          dont
r   O L        d Sent Self-Awareness 4 @ aGe Ø @ nØ I nose I-self
a        System        (i) yet, I onely Nose what I = told,        L
Z-bRa   i        2 concept of «i» D.N.3.        feed  whee|HoUS3
T   o   e        1 e   6aSK        Just MOM + DAD        1 e t        S
naught (ZiP)   > Ø = 1 Pin Point on ^ graduated (6inary) Scale fROM witch
e   One        good + Live = wRecKoned (dead On A rival.
r   N (Ø,Ø)        out of Print.   6 (cyanide) Y        =1Ø5Ø [□]
```

                    T E N E T

                    Ø . Ø

1   [●]   > i.e. 1Ø-ET Ø.Ø of
Ø   (in ^ vaccum)   (mere img of Ø.Ø)   (iRL DiiV)
t   ∩EST of 1ØS, 1-off TXT ware (Ø, Ø) = origi∩U|| aerɘa ƆODƎ w-
here-in U (in8 iD) = 6orn mm/dd/66 , + FT 1Ø-ant teL 1Ø ± 1 yr) + if 1
       types ^ verticaL Line: «(Ø, Ø) tideꙅ revɘrꙅed on 8 set animaL
   ɔode doȼent   R
       i | peeLꙅ» 6aɔkwoodꙅ U gɘT «s⅃eep Laminateꙅ (8 nodes) re-
       VerSed iT (Ø, Ø)» 5tarTing w/ i-ba||s, then i (authori-T (2 ^ T)
+ D   i C i = 1   here-in) evolveD X bi-iLLogicaL (r&om) dubbed bOd-E
O   d Ran   U   type 2 movE 5aid i-6a||s a-   -
X TS   e bY   V = 5 ЯRoU∩d X L&scape 2 Sea + [●] 4 1-SE⅃F (ꟻLƎ5HED   e   b
       P   OoLogicaL
iS   T out) > 4 ∩ ∩tire year we Я ∀LL age zerØ 6ut dont   ∩   g   s
          o   (doG 4bid) Ǝggknow|edge «Ø» az mile5tOne, fee-
       Ring «zerØ» az U∩Holey 5pheric Void, ∩U|| 5et > 5um ceLebR8 1st d
       e   moonth 1 or 2 w/ c&Led cakEs, u dont REM-   k   o
5eLf   o   R   em6er + your paRent5 ≠ aLive 2 a5k   eL   impromptU
       eYe6aLL   o   even if they Sti|| = kicking u wood naught   ∩
       r theory m   tRu5t their takE on what happ-   i
       e   G   a   end > reꙅiduaL D-Sire re5ides i∩ DNA + «A∩D MA i 4
       nu||ify   nØ concept of Ø   O   n   M
       c   1 ⅃ᴉVƎ ꙅ ƎMᴉT 1\S ⅃OOⱭS ꟻ⅃OW H-ꙅƎ⅃ꟻ 4 TᴉⱭƎꙅ ЯƎVƎЯꙅƎⱭ   o
D-serve Ø   o   O∩ 8 ꙅƎT LAMᴉ∩A ꙅƎ⅃ꟼ» = 1st Line
D = 5ØØ   m   1 = i
decimaL   in Prior 6ook (iSB∩: 978-1-94Ø853-22-2) or La5t Line iF   ∩
w n   o   d   i Te   T-ick
e   red 6aɔkwords + ¿now Used 2 Seed z next? in keeping z mom-   iT
y B   i   e   n   Bed -Entum (¿Dog ꙅɘɘꙅ oⱭ¿) yet @ tH
    +   s h   reaLi5tic   5ame Time 5hake o|d haBiTS + ever-pre- E
b + W, coPied?   p   o   dont
r   O L   d 5ent 5elf-Awareneꙅꙅ 4 @ aGe Ø nØ 1 nose 1-se⅃F
a   5y5tem   (i) yet, 1 oneLy ∩ose what 1 = tOLd,   L
Z-bRa   i   z concept of «i» D.∩.Ǝ.   feed   whee|Hou5E
e o   T   ju5t MOM + ⱭAD   6aɔk   e i   S
naught (ZiP) > Ø = 1 Pin Point on ^ gradu8ed (6inary) 5caLe fROM witcH
  e   One   gooⱭ + ɘvi⅃ = wReckoned (dead On A rivaL)
  r   ∩ (Ø,Ø)   out of Print   G   (cyanide) Y   =1Ø5Ø [■]

                    T E N E T

                       1 . 0

        n   ^   K   ^   n
[c] 'PRoNoUN': «oNe PoiNt Zero» 1.e. 11'th 1Q-ET   /\   e
1 2 PHrase «once upon ^ time» = moSt commUn waY 2 STarT ^ bOok
1 == 1 on 2 dot: 4 good reazoN; make⁵ SeNse; nQ? revitriNe 1N
1 = «1st PersoN 1Q5e»   oNce upon ^ time 1 goog|eD Gada
+ «2nd PursoN» = U 2 reader   2 fiNed out whO 1st iNvEnt3D < s
1Dea of UNity; how N-E # multipli3D bY 1 |/
eQui||s saME # baSk (aka «1DeNtity 3Lemen't» > 1st softWhERE v 4 ≠ nul|
h (OR OS (opera8ing SyStem) VerSion = 1.Q; 1st itera8ion; e
no LoNger in Beta tho 1.Q DNA (≠ 3X15t)   1   G f
iF U fo||ow 2-noSe errorow U w1|| nvA aRRive 2 3NG rt. G ^ r1gHt angle
>> = ∴.   T ∠ (90°)
1x up on 1 time 1 (= 1 = Cal A. Mar1) 1sSued zzz ><(('°)'> iSsue Q d
thaN 0.5; 0.75; 0.875 + 0.9375; NeveR ReaChing 1.Q   0 0 0 0 1 e   Y
1 m o   = 0.1 in Reverse | 1 0 0 | n
paiR-0-docks   a   smorbriLag of fDeNtity matRfX = | 0 1 0 | nQnLinear
Time u   witch mu|tip|fed bf NE VaL-U = VaL=U | 0 0 1 | a m
1 itseLf; tho octo-n ha|v3 111 <3'⁵   T   1 0 0 0 0   Y 1
K-05 + 1X CPU⁵ > «Linear» means ^ DynamicaL SyStem dat can b SoLved eX-
act|Y (az opposed 2 nQn-Linear (1.e. K-05))   eLonG8ed   1 r
> 1.Q = Probabi|1-T ^ eveNT = 1QQ% certain 2 occur   m e 0
e Knee-high > theRa'S ^ Neon 0Ne in Lone|Y + «2 can B az bad az 1»
sang HaRRy Ni|zson   a 2 Lone|test # SinCe z # 1   o
+ «nQbody» = name 0dYsseus uzed   O   m n
2 get awaY from z 1-eYed cyc|ops whan aSked name/ +1 = 1NTL DiaLing coDe
4 US + 1 = true in machíne coda (+ 0 = fa|se)   AutomB   tA da
0; ware 0 + # = # itSeLf : LaNe 1   > X foNt = «SorCe CoDe»
FoLio 2 PorT.   V   + said Jrogramming LaNeGaugE 1 tYPe in rAw
bfFurc8⁵ a   A 1 1-of-a-kind (each 0BJect) = UNicode text   1
Y; SidewaYz = X1nG   + f = U Scan in 6ars af ^ L8er D8   n
tab|E p   J1V3 2 V   V = ////   nvr Dye in agony Unt1l now   E   h
each Thief >> So2 2 reV1Ve + home Technique 2 0bfusc8 1 o GUAC-A-MoLe
karmA W   V   o   up 1   A   + ELiTe 8 1 T E r
saP nOD 7 X⁵ Offeeat 61rd D-fies Zoo⁵ Logic + init18⁵ e e WORMHOLE 1
p1ñe E R S|acK X X   e   f   e 3m6ed 1/0 coUntdown a saga G8 MaKe T
b1Rd P1b Soup 0 F   X = ////   RefLection seekWinds re d Y u   9
TryanGuL8 X X-terming w/ nQ decaheDrA rebo0 T in se-ment   9
                UNiVerSaL 1-wORD [■]

Ø . 1

```
SLiP          A [●]                       Po5itive (+)
T                                or negativE (-) #s > iT takeS 2 2 tangO But
R        How dib the5e 1st 2 come in2 Exis-1Ø-ce? «i + i survive» sAng H-
E     ¦         R (of B²) > in sum circ-L-s, Ø = «cipher» U5ed 2 D-cipHer U
ATOMs --¦       i  aerɘaL $Leep PatternS from deatH > 2 B 6orne A- n      MU∏
M  A        D    1 proTon                Ø⊥ dɔɒʍwoɿbƨ = X      ⊓ F
   DAM, i = G    e|ectRon      wake = 2 «zerØ out z 6ookS»    ∀naLOG
          EDE∏          i  5pot                a       (∞ Z-rØwS) + U
∏Ø BA∏∀∏AS               ∩      re5eT coU∏ter 2 Ø.ØØØØØØØØ   MiiX = 2ØØ9
O          > az ^ diGit, Ø = PLacE-        a                   O e
D  Hi/L      4                Ho1der 4 ∩on-Ø #s, diG iT? an∅n i'M Us = U ± i
   A  O   ± st&S 4 sum 1 (or thing) E|se dat comeS 2 LiVƎ  m          ∩   =
O  ∩         az X-tra-terrestiaL 1Ø-iT @ X,Y = GroU∩d zerØ  P         D  |
F  G         k    m    e t                     Lo9ica|     r          O
FRUiT        e    e    vira1  ((  Ø.1  ))              Diarie5
   ∩         ∩ pLace     i a    G                   + 1 = i   n e   X =
2  G                s c      a  ZerØ ini8S w/ z La5t Letter
           (in αLpHaβetic LanguagE) 6ut z ∏Umeric    t    d       ⌘
«d1p Ø.Ø» (Ø:ØØ)      &   s  i  9  Repre5entation    O            +
   r          (      w    enveLope (Ø) = 1st #               )    Z + C
in o               r                                             bTn
   m       > αLphaβetic 1 5tartS w/ O, 2 5tartS w/ T, 3 5tartS w/      R
∞  Prop       u    T,4   "    " F, 5 5tartS w/         U a          L
   t i        c    -           F, 6  "   " S,       aBove
L  e p        i    1 7 5tartS w/ 5,                   | e 4m
O (di5enteGr8eD)   Ø 8  "    " E + 9 5tartS w/ ∩        Z  (Ø° K)
O               > if U Take the LetterS (∀LL 7ƎVE∩)
P|aceHo|der tXt                                      R
S         {Z, O, T, F, S, E, ∩ } + Rearrange them / U 9et «SOFT ZE∏»
 C    > if U drop Z (b/c Ø doezn°t Exiʑt in nature) +
 A                      A||ow dup|icate LetterS, then 1 can
  SpeLL the titular «∏E5T OF TE∏5», ReaLizing z initi8ing TE∏ƎT Ø.Ø:
T         U = 2    «∏ET 4 TƎ∩» in Reverʑe = 1 + z 5ame
   L         (tho «anagram» ≠ Anagram 6ackwoodS, 9o figure)
neT = red i > in Σum, therƎ Xi5tS ∩ inFi∩ite # of #S 6ut they a|| B-
  O   in 5pa∩.                  Gin w/ 6:              h      T
  ∩        E               F-E-Z-S-∩-O-T + ∀LL #S end in      Way
  E-O-R-X-∩-T-Y-D, witcH rearrangeD = TE∩ DRY OX, bwayʑ [■]
```

TENET

2.0

T Ξ ∩ Ξ T

Ø . 2

[•] (in key of ii (in8 iD))†   > ∀∩D (+) 2 thi∩k reQuires ink, i∩c.
   ii (i i i i) 5in9 (≠ Fo1kLore) i∩ the BE9inni∩9, 6y chants on hOr5e
      ∩ayi∩' 6ack 4 crYin9 Out LouD in rEBirthi∩9 ritua L   M
∩U∩ ha6its 4med, n∅ wives tai L (in yr LoGBook) nau9ht that u [•] A 2n,
   2 z cOntrary con5ider HumU∩, (PriMe 5crEE fLow) = WOL˥   D
5Leepin9 in 5Heep cLotHes   (pAjama/T-5hirt)   wiLe pa jams ^ ban-
ana an59ram 2 6an fruit of tHe LOOM   ∩U∩ ∩one 2 me @ z time
   (that ii = a wear of) n∅ 5hoes n∅ 5hirt n∅ sErvice-i/O, ware i/O = ∅,
Combin8ion of in- + Out-Put wHen 1 5Leeps (w/ 5wayi∩9 ñose) U∩der voLcan
O   -O, ∅ (on z 5ide) D-rive dada5trEam in B-day 5uit @ n AGE b4 1 com-
M   π-1es ^ ñosE (prim8) 2 primE pump @ GR∩D zer∅   G
E   (i.e. «i Point zer∅») = 1 uva kind 6ook OBJet
TALE (nau9ht 4 $ai L), 1st 6ook (of 4) of 4ier X-4ms 6y ∩o One (iSB∩   A
   ∩   a   p 978-1-94∅853-12-3) ≈ p&emic LOG/Liebury/jour∩ihiLi5M/
   Durin9 x   e   9urney wHen
'66   L   c   wHere Dayz ha1ve n∅ #s)   (in REM)
  M   e   i   + nites = sonAred w/ SOU∩d ∩aviG8in9 + RAn9in9
SHOWER   s   a   fROM a9e ∅ 2 1, b4 uu = 6orn Ma 5Lept on
P   A R-me surPLu5 cau9hT   fo1dEd out
E   T   s   (nayV 6Lue)   i∩ z Livin9 rm, 29 yrs b4
LEO∩iD   U waiL meat ∩ «R-me Of me» (1995)   29 yrs b4 2day (2/4
L   E   a   wHen cur5er/proMpt fLa5Hes, in ^ stairi/O   /24)
   Re:purpiss 6Lue (4∅∅∅ Å) hue 2 eAt Leftover TV din bi/w/ U
      (fewchair ver5ion thereof) hiDin9 in5ide ∩AYVEL oran9e, pudDin9 U∩-
eaten 6its in subzer∅ fRid9e   2 L8er r9Hash in2 kit-1∅-PA soup
            t   ± ⊃AT TACOs = 1∅ET ∅.1 (+ MA SLEEP ≈ DAM nO-
MAD (anti-DrU∩K, ma) of soLe kitscHe∩, nau9ht privy 2 drama B-hind A
   n   O   a   cLosed Door
   i   O   i   #∅.1 > 8iRD RiB < 1.∅ #   n∅ adS
   nau9ht 6in LAiD yet   + vice-ver5A ∅ (°C) = pt @ witch H_2O
      LiQuid turns 2 5oLid > z duM6w8eR ∩-ab|es Quantum tU∩∩eLin9 B-
tween D-ba5ement + Kit5cHen 2 D-LivEr muSh-Room 5trokin9off, akin 2 hoW
subuey ∩-ables 1 2 tRaveL from pt A - 2 B   ∩   H
   c   i   r   e   w/o Cin9 z L&scape, abOri9i∩UL rEar-   Y
ran9ement of «M-E-D-i-A T-E-R-M», Ox, w/o 5kippin9 1 6eat, 6i-pA55in9
   aeriaL   e d   M-T 5pace (7th «Heaven») thou Rt in [■]

---

†inhaiLin9 HeLium

TENET

T Ξ ∩ Ξ T

Ø . 3

    [•] te5ting 1, 2, 3, D-pre55 D-Lay petaL / 5et @ 1/e = Ø.3678794
Ɛx   5eɔondＳ 2 Achieve Log-rhythmic DK, ware e = X-
 ponentia| + e^{iπ} + 1 = Ø = enough 2 Make yr CPU X-p|ode in2   c e
po-1Ø-tia| hi ponytale, π = pi(e) + i = √-1, ɀo-ca||ed imaginary # /o r
 ɀ ∩Um6er z Bedder / more com4table ∩Um6, Loose 9i||Ｓ 2 Leaf C + re/pro
 e   E   -cre8 ware 8 = Ɛ + 3eee / 2 Bee or ≠   y x
3rd time = chArm (thx 2 Lithium)   T   (,)
coU∩t of (o∩-T-V) 1, 2, 3, 2 ｐRime z Pump, on 1 of tho5e   t   w
 n   i   «U R herƐ» MAPＺ, U R @ GR∩D Z-row, where Z =
        K verticaL aXi5 in 3D pLanƐ ----------------┊   i   e
        > 5tory of yr Life (4ever Living @ GR∩D Z_o) ┊   n 1/3ee = Ø.333
      A-peeaLＳ 2 innersent By-st&erＳ st&ing Aro∪∩d ┊ aLPine e
 h&Ｓ in Pocket w8ing 4 apocaLYpse 2 e-Laps   R   ┊   t oui? Back 2
  R   P   Ɛ||ipɀis,   2 B-Leaf Ɛ-Den ≠ ┈ 9Ye !   A
LOOPing LapＳ ware doG XistＳ herein 4 A|| iii E   s   v   ! YeɀɀiR
  i   E (∀ i, ware B-A-C = mnemoniC D-   ∩Ø ｐarking e 3! = 6   ii
  D   vice 2 REMem6er, 2 9et 6aɔK 2 ware   n   P
   3   π   A   wax 3 s&wiTched B-twee∩ d b
   D   i   1x i B-Longed +   A   r E i y   2
      on CoLo∩ (in 2004) i w8ed 4 the BC/AD (Bodh[i] CircU[it]Ｓ /
A19[a]e[bra] D[ra[in]]   R   t R   WaveforM
L   (iSB∩ Ø-9746053-5-2)) + on Bdwy U can ketCh 1/2/3 d   LOGic O 4 P
Going U∩dergro∪∩d   trainＳ   R   3
E   b4 «On Broadway» = ven-U i 1st cot mu5ic Live, Down
B   z 5treet from 1st top1eɀs (1964) + then 6ottum|esɀ (1969) rip c|ub
R   ware ^ 6oU∩cer dyed having inner-Cour5e w/ ^ 9o-9o dancer aFter hrＳ on
A 9r& Piano b/c tHey axidenta||y hit O∩ + the autom8ed ri5ing piano rose
    O∩ iT's onE   nip-Ls Lit   i∩ 2 z c:Ling   S
    O fU∩ctionaL   trapping z Coup1e + the 6oU∩cer = aspHyxi8ed (DOA)
   wiLe the 5triPper 5urvived b/c 5hƐ = thinner y   o ii   ii   nn
C   v   + ∩bɀɟds-qoɯʍ i (!) = factoreaL / mu|tip|icaTiVe   o
ooZe e   sum uv pre-5eeding VaL-UＳ + bi 5ame token the 5quare root
n Of mi∩Us 1 = death bi drowning aLivƐ on bed9viled bed n   L   8
d ∩ t h   (same 6eve||ed db u'D Lose yr innersense 13 yrＳ L8er)
o EnO   a ∴ u, 4 octo-π ha|ve 3 <3Ｓ   n   en   p
r R   d = i d code 4Ø8 = ∩Um6er One (1986) = aL6um that maid me reeL-
  8   eYeＳ u Live @ 9rnd zerØ > 2 «Fee1 MT in5ide» = 9ood thing, nØ? a
  S   2 cOme in 3Ｓ 9   ?   !   b4 U/i DK 2 de6ris   db Lay [■]

T E N E T

θ . μ

404 / nutes Y,X unbound / dayzz +

out w/ nowhere 2 go

o/ink on skin   o/ink Y n0t?

what agreement can u make w/ Yr self?   gold font

onely that u'll reach the X-iste   Extraterrestrial Y

ʷ/ kin in knots   This erroria

useful Y   anoDal   necessity = num

orbit   arching   in ore-bit ^round

wayword sun   untie   Yarrow Stalk

Sphynx topotype

heXus 6 eXes X X X X X X X·X X X X X X X X X

lig† eYe t   iti t eYe t

wane †   sheer windy   undefined   Ybniw reehs   2 B A-ware of dice

non plus parallel   b-come untangled   b-come untangled

Try 2 come BH U arRive   @ Youth   axis

go 2 BH U h6e†   FLOOD ≠ RIVER =   abacus

vertical

404

TrYangul8   i   r 18 H

Δ iz 2 process in CPU   iz crou flies above

Old u8 Dk9 2 ǝuɔ 2 reHn ^HetHer Yolk thesis   e loony

teknowlǝdge D-solveɔ 2   iNHu obeY majd i]   step f(x)

Ø . 4

```
[●] > 1Ø hut! pooL U-ee + come face-2-face w/ ^ reLick of yr 4meR 5eLf
btw, bway,        BeRRY||ium                          A        iV E    E
    R   2 4m re4muL8ed                  dowՈ bdwy UՈ-      Ո R      Ո E A
    O   4                  m   c  Ti1 U reach an aGrEEmEnt w/ 1-5eLf (Few-
    A chair 4m of), wherein u also ReeL-          O    |
    D     EYƎS Ո-E word kin mene ՈE-Thing eLse dEepenDing on contXt, in-
Tend S? 2 ReaCh X reaLization makeS U nod in UՈiSon/cOn5enseS          a
R  e          c  w/ iV in Rm   2 concuR, nØt 2 Ոod off, ox > 2 in-1Ød = 2
i  a          X-haiL    UՈderst& ∀fter, Entiendes?  4in 2 meme  d    1Ø-4
GrrrrS 1 B-9 K9 (pick of the Litter) (4m of ƖiTerasure)    rubbrr meats
          4       1 in Bu5h = eL pLaceR (coUՈt-E), caLi4nia   c   ode  Q
  4-in  4M           iV   ±          U = 6read frOm 49er 5tock 1st +   u
4mosT, 4 4 every 1Ø iՈ 1 rut U Loop ՈUBi1e ЯAT2 BaɔkwooDS 2 5PooL sin-a-
      O              meՈ bUՈS 2 sՈUb ouT Bonfire + Jack ^ Mac/trUck Re:   r
4muL8ed fROM 9o1d-LEaf $uper $tringS --------- E          L   O    e
  L-ectric        R              , GroUՈd in CharcoaL in2 LampS
          from Tenet Ø.4          U 5hoot @ Tiɔsue Ρaper        Ex-1Ø-d
iSɔuing 4th ≈ G-knee w/ ^ foUՈtain PeՈ  (eQuiLL 2 ^ pUՈt)
iՈ 61ue-9ened BeLL 6ottumS 5aying «O, pen Sayɔ mƎmE, ^ Lad in5aՈe»   LL
(U = rUՈt)              2 mine «Ore ZonE Z TeՈɘT» + re4m Z-nOse paira
2Be           1Ø-iT witch st8S U wiLL nvr rEach ½ weigh 2 B-9in W/       d
  UՈLeɔs U ri5k sEwer5ide bUՈt              (sQueeze pLay)               o
    o     w/ 1Ø (iՈ turn) in 9 rutS U Loop ՈUbiLe, o say? 5trut 2 home  c
    s pL8           C 4 yrself   |         (iV in Rm)                    k
  wave4m   4       (H) decked in TooLS of i9norance i LooteD            s
@                    from DAƆ's Urn in 5ink w/ dbased        E
    ± Ø               rub-Being, ReUՈion w/ prOgenie         A
^       s-π-der       b        i ՈevEr ՈU Ǝ-XiStEd in 1st pLace
   saw-2th    ∴ add «in bed»                 A        a r
a   i   o    2 every 4tUՈe cookie, 2 Augur OЯO ± oRe-O 1 AughT nAughT
9et knickerS in knotS + kidnap             yr oՈe Kin   2 4in PLacE
ƎBƎB m e U                    Re: caPer io      ?     L D  e  Ո
 B   pr L  Ma pooLed off when i = a9E Ø, O    y         z TiME       Ǝ
w     omp      n    i o oVer  4 yr oՈe 9ood H     h Ո         T
hayZ t    1 eLse       L  u       Ma 5aiD        E    BrokEn
e a Au bar       up 2  L& S-caped      (bi 4ce) in X-isLe m
nu99et foo|ed (π-rite)    e     a ՈU 1 in ^ 4-in st8 Lswhere [■]
```

TENET

5.0

□ = t y e s
e = Limb

Paradox led 2 entanglement

'94 cyberspace vox on Agora

R knotted nets meshed in2 1

dada salving in2 caved babe

Liken 1-self 2 [Chen

the day time stood still

Everywhere @ once penetr&8s

arranged in complete Shell

disseminate

S A T O R   paper

obfuscate replica   adapter

P E R A   tic b   outcomes

everlasting

lifespan allocate   T E N E T

eXperimentation

always   A R E P O

halfway i c mere

rearview   R O T A S   R arrived i4 inquirer

Reading huck fun&Gaine waxe

still Live in the innertacE

Vacuum up the vacuum itself

circles R never complete

Reset TENET's packed deck

50/50 + U keepS

Embodiment of user X-face

mind\body dilemma reconcileD

T Ξ Π Ǝ T

Ø . 5

[●] > Ø.5 = ½ = «1 + 4 a + Ǝ-ViL si ʇoɹʇ / a man si a sign > iT's
∀LL ⅃eft \ «suDden words aɹe torn in age» / 6egan ^ 5tory / ^ Daʏ ROTS
\ an Age 6ɘɡan i∩ ЯOT / Eras drowneb .... du5t feLL, La5ting ıs A sin
      R   A ½ Life   r ^ Man rots / i ⅃iVe + a 4 + I» = 5.Ø = V
C > rE5et Tones   > nay, kid, ^pe = animaL, FYi   S
H   A man a pLan / fLy Que? pa∩ Am acroƨse L-5haped i5thmii 2
EnLighten i in z proce55,   4 eyɘs = «||||» 6ut 5 ᗡ.∩.Ǝ nVr
C Ve5tigiaL ∩ET 5kinned from ALLiG8oR   in ha5h marx .. on P-chEE 1 can
Key = B ∝ WhitE in 5/7 ratio on pianO   cheat + rite az «⌿⌿» aD-LiB 6ut
       i write X Opera of U∩icode texT   i 5wore i wdn°t u5e iMGS, btw
   «卌» Ǝxi5Ts (U+534C .... tran5L8 2 Chine5e 5ymBo| 4 «4Ø», in facT)
+ getting down 2 bra55 taxes (fi|ed) nØ 2-D 5 x 5 po||endrone ǝxisTs in-
g|ish, 6ut in Latin there's the S A T O R 5quare, witch ties in2 1Ø-ET Ø.5
whi|e i 8 U∩a arepa reaped from A R E P O, z farmer who 6ui|t ^ temp-L 2
C: witch doG wood 5how up 2 B ^ T E ∩ E T in the hack 1Ø-t + z doG of OD-
D JoBs ... St&-in/dub-L in text O P E R A, of peda|s in BaLoom that 1ead 2
rot, 5hort-⅃iVeD nada, 5pinning R O T A S 5pU∩ from Gen-e's 5pind|es 2 eVe
-∩ the 5corE whi1e E-ting Cheat-   -Os   (1Ø rU∩s in z 1Øth)
         D-Ba5ed 1Ø in zhe LiBrettO ware z tenor atoned, often off 6y
 as∍-iD   1Ø in ne5ted LoopS, PooLed bi ^ mu1e (U∩i5ex donkey) 2 hi-
   o   jack ^ error p|ane 2 Try makE ¢ents of L8er on 5cense.com, 2 re-
   cre8 ^ 5cene ware DAᗡ kidnapT i in Return, naught 2 LU∩done nor ToykeYO
4 tat 2   n   6utt DizzykneeL& / L.A. + DAᗡ tries 2 di5track U   e
   T   t   bi h&ing u crAyoLa- (fLipping tray-tab|e down)
                         crayons (8 of 'em ... 2 4m 1 bite) TeLLing U 2
draw pink L-if-ant5   U   dat L8er u tran5L8ed 2 Vi
 (ViƨuaL editor in U∩iX OS) > ½ the Time u don°t nØ what 2 think ≈ X-rays
  in retrOSpect u tore yr one heart Out + 8 iT, terroring in2 fLe5h w/o
  a bee BLeeding thru 2 the edge, faR from Canned-sis (dotter ᗡAD
nVr   ∩eVer eVer had)   anD e c i m8 = 2
say Letters in «A DECiMAL POi∩T»   m   T
 m   1 can rearrange 2 4m «I'M A DOT i∩ PLACE» i   i   p
 P-2--P   r   z deciMAL ▐Pt Mid-weigh B-tween 5 + zerØ, Bway a Ǝ
 i taiL   BoRon   where 5iVe = V in Lati∩
RE-Vamped +   VirUs   + «⌿⌿» in ta||y mark 4mu|a,   t   h D-
-Rippped from iV in ∩ improV LiSTSERV   t   V   [■]

GHOSTS 6IX FT UNDER

                    T Ξ ∩ Ξ T
                     Ø . 6

6, 5, 4, 3, 2, 1, Ø, ... [●] roLL z 6-sided die + get Snake i's ⊡ ⊡
     Honing pidgin zeRØs in in para||e| 2 D-Liver in X, Y, Z p|ane |
  (O,O)   dominO pizza π   ($auce-age + mu5h- R   hoBo
vrrOoM) 2 Hangar 1 West in 66 V6 musT-   U   hippOcampus
   T E   9Arden of 6a66Le on when z Rite 6a||joint $naP-   X-r8ed
h   ∩   Ped, 5ending i in2 ^ 5piraLine 5kY   v 5peCs
o Zip-aLiGning in2 dArk 5ide uv moon w/ $C_6H_{12}O_6$ in my   weather 2 Ab&don
me   T p|ane dOminOes w/ Gr&dad's 9(H)o5t   i   Rai|5ide
o Mixing 1Ø up + picS [▦] ti|e   ware U a55ign 1 MOLE num6eЯ   (卌
n tO 6 x 10^{23} iter8ions of the f-ing Ho|e ware f menes 5tioK ^ 5tick   =
intO GR∩D dig-iTS   > LiVid E|efants Go BoUncing E   9||)
Down FrEEwaze = how i rem-em6er pLanX Con5tant, ħ   a
   (+ F-A-C-E = 5paces in B-tween   (+ Good BurritOS Don°t Fa||
Vizual eDitor (WYSiWYG)   Apart 4 z ba$$ ¢Lef))   e   U   e
+ Eat A Dead Goat B4 EasTer or   a   T   v
     S   E|efants And Donkeys Grow Big Ears   d   Ǝxis-1Ø-nce
CAGTUCT   2 reCa|| $trings over f Ho|e   r |
a   L R   tho usuaL-E u tUne tO D-A-D-A-A-D sew u drown out z 1st
r AsTrOphy6   + Stick z tonsi||ed mouth in2 2nd
b i   caVi-T's ear 2 4MuL8 ^ Thirdrd tongue 5th   L
o M   ware eaCh iter8ion CompoUndS z preVious 1 (w/ 5peCuL8ed intResT)
n A + @ once decon5trux   p w/ freakwindsea 1/f R
             z proLifer8ing 6-foot graves that manifE5t R-   toY
bit-rare|y az «Lamprey» jeans witch in ^ pinch Can B ru66ed Sidewaze   A
@ ang|e f° 2 give rise (X-amining z X-rayz)   t   2 ^ f-hole   ∩
guitar Carved from ^ 66 VVV-uCk|e   then p|anted in 5oiL—   G
   L po5ting C1aim 2 4niC8 w/ 9olden F1eeCe (aura||y)   U
             dont di$coUNT b|ack 5heep e (i-very + e6ony)   L
Com&S domin8tricks (mu5e #6 w/ 8 RmS)   w/ drum styx   8
 r   o   2 serenade U, singing ^ 2n (^tUN Ø.6):
 9   we |ive in z Craw| 5pace 6-tween z 5heetS b
 a   d   patterned w/ 6-sideD 5tarS 2 bee bedber in debt (5Leep-
 n   ping in z 6th 2n ≈ z 3rd) 2 honE in on home w/ pi||owS bi||owing in
 i   turn 2 4m ^ Cumulus Virga b/c Rome ≠ 6ui1t in 1 eve, bi ChantS: s
 s   a|| ye   a|| ye   oxen free   e   LeG
4ming Cry5ta||ined Co||ective fROM ^ me55 of ivOry + CAR6on   x
                     [■]

Ø . 7

[●] 1Ø-ET Ø.7 = 4muLA 4 how 2 cook se7en seize soup > one up
on moonday 4 ∠u5t over ∠ucky # 7   > 2sday atone 4 G|ut oƒ ∀LL U 8 z
weak b4 + 4MUL8 pLa∩ 9oin9 FWD 2 Estim8 + ma∩ed9e eXpect8ionˢ > Re- o
ƒa5t[»]   5et A|timeter 2 ØØ.7 >   ∩
         > do nau9t in5ert Tongue yet > z 2ⁿᵈ 3ʳᵈ drownˢ out e
z 1ˢᵗ 3 ho||ow-bodies (6e5t ∠aid)   tho n°t B4 z 1ˢᵗ, 3ʳᵈ + 2ⁿᵈ
cre8 ^ 5ᵗʰ ƒ-hoLed cavity 2 receive ∩oize > a9ree 2 n°t B 9reedy
         > don°t B tempted 6y 7-yr itch > z 5ᵗʰ 4mˢ ^ chord w/ z 1ˢᵗ +
3ʳᵈ uddering ^ 5o∪∩d B-yond what 1 Tongue cou|d ƒa5hion on itˢ o∩e
         a'cord > z 'i' in 9enie 9etˢ Rea∠si9ned 2 oversea a∩ima evOLutio∩
> in5ert Leap day # 7/11-eve∩ after OX-ƎYE madru9ada (½-way) + raP up
 1Ø-itus (ring in yearˢ) > i 6een 2 Gr& Caño∩ tho n°t Gr8 wa|| > U C 2 +
3-toed 51othˢ + 5tick tongue in ∩M∩E (w/ 1 i∩/Out ho|E) where C ∩M∩E =
5∪Bset oƒ LampRey (w/ ^ FWD hoLe 4 i∩ + ^ rear ho|e 4 Out
 A   i/O ≠ Ø | 4 ƒ-hoLe di99inG ≈ moLe) i
[SiC] > oƒ #ˢ 1-1Ø, z mo5t comm∪∩ Fave # = 7 > ^ wRath   ∩-1Ø-ded
 E   > K-OS reiter8ˢ di5cord i∩ 5itu > A∩TarticA is ∠∪ЯRou∩ded
 bi ARCtic C > iF U take sum oƒ 2 X 6-sided dice, z Pro6a6i1ity =
hiest 2 ro|| 7 > iT took 6 dayz 2 cr8 ∟ m∪∩do from 7 Hi||ˢ oƒ Roma
 + Oppo5it 5ideˢ oƒ 6-sided dice (indus-tree (+ 1 daY reSt)   moRe
st&arD) add up 2 7 > dont 9ive in 2 ∩V (Leadˢ 2 DK)   a
 e   d   > i 1Ø-d 2 Live @ Gro∪∩d Ø > ∩eutraL pH ba|ance = 7 o
in «Ed9e oƒ 17» i thi∩K ∩iX 5in9ˢ: a   b
«Just Like the 1-win9ed dove» + eatˢ 5hark ƒi∩ souP > 7 tick|eˢ   5 e
noteˢ in st&arD 5ca1e: A, B, C, D, E, F, G o a A   h   decay
  m 5kate   R   i ∀LL se7en continentˢ Lap   turista a
 map   e   A 5tart w/ A eXcept Europe c u A   c   o   ∩ature
  LaminaR FLOW w/ 6° oƒ Seperatio∩ 2 D-con5trucT   o6soLe5cence   L
  a e ∩   E   o   r   ∩ d i L a a e a
  s L a   R   k 7even wonderˢ   SLo a∩o∩yₘ o r n   s
mo5t dupLic8ed hou5e = 77Ø   a   E9yptianˢ 6y O °u∠ 9r8 pyramidˢ t
 dis- o r had ^ 5ymbo1 4 zerØ in Acco∪∩tin9 177Ø BC O o   o a a i
5tance G u t p r   e n [sic]| e a &   w a USed 2 i∩dic8 tack
 r   d-base LVL e   α|phaßet Soap6ox ƒo∪∩dation eco   c e h i y
mea∠ured e   h anima   o r Oƒƒ6eat u   L v u w picchU 8
re1ative 2 Oñion n ∩ k   L-bow tofU rs   dbasement   a   o
  p e tocK wa|k E ƒa5t   s   Pride   s   o   year∩ [■]

com.

learned 2 uze computer5 in the '80s

by 2010 they learned 2 uze US

memorex

Eruption of Mt St Helens

start of DIY/Startup

rhythm

when X, if incubt   Lennon got assassinated

nostalgia   Y-Hk u

Pel'd groove a   threshold murder   net5

eighth

+CD

80% of FX

gone from 20% of cauze5   D-Line8 orbit

n0thing

Raygun

last o

turbojet

number

list

knock

link

J Pablo II

e90

800 8

Visas

VISA para dirigirse a MEXICO en calidad de
VALIDA por
Num. 336
Nogales Arizona
AUG 11 1980
PRO-DEL CONSUL
MARGARITA MANRIQUEZ
CONSULADO DE MEXICO
NOGALES ARIZONA

abogado   avvocato

SERVICIO EXTERIOR MEXICANO
VISTO para dirigirse a Mexico
CONSULADO DE MEXICO
SAN JOSE, CALIFORNIA
AUG - 7 1980
PRO-DEL CONSUL

T Ξ Π Ξ T

Ø . 8

        R
     PRE2Z   [●] L& Octo-π (taRanTuLAˢ) =   6 x 8 x do   mi fa
so La   S   T   in8 in my iD B/c SLeep LaMin8ˢ 8 nodeS RE- do
    ver5Ed 4 Lack of Oxygen makeˢ U DanCe in hy5terix ...   TMi
     sumThing U 8? sπder takes ^ 6ytE «Dont coUNt oN iT»—magic 8-Ba||
    U   4 a|| in-1Ø-tˢ + puRpo5eS wE (az ^ hearD, @ $t_Ø$ = Ø)   B
caN @ be5t UNder5t& R One 5pecies   iF U = ca||ed upon 2 5peak O
typE writ-1Ø   con5u1t coPLaneR eG96is-1Ø-ce: mine repLy = NØ @ $i_Ø$
OctaVe (8ve)   U may reLie on tyPefacE :   2 OcQu-π head 5pace
    E   uv V8 iNgene [ ]   LeftoVer   in M-π-re st8 + m8 w/
tRaNSisteR i   ● REDorder Eggˢ Ǝxi5t   b4 chicken checkˢ out   R
 A. i.   ULtraVioLet vestigiaL   ЯeeL-2-ReaL-ay ^ Lo5t PineaL GL& h
 D ReVerB   E Po55e2sinG t FLuxuS   a + faciLit8 nap st8   Diy
 i E i   R   mi55ing LinX   Every evening   SLeep ea   pRim8
COPiNg MeCHaniSM 2 Po5tpone on 1 siDe inevita61e T, Time ch 3Ɛ   O e
 H   i   o iT = D-sided 8:5   e   v en pin E ^ L   3Ɛ mi2s
 eYe   ThreadinG d a   whO e   e dormANT arX w/ 8 x S D
 6   «iT i2 ceRtain» d   4 ^ eoN rock   e i   U
     G .   8 data RiveЯ ɔLuster   pinea L   4 B = 1, 2,
3, 5, 8, ... ∞ a   i   2 anNUN5i8 in-1Ø-t outriTe in 8-ba|| Ma55aGe
    e   derived   c   6an r E e   A   8 s
FibOna-Xi # ♭ da a e   ♯ key in Sorce accumuL8eD intO amoUNtS 2 W h
    s   pLot   e w n   Ever endˢ a   s X bed A TabuLAte
Shared via direct c w «RepLy hazy, Try again L8er» s B teaR e Y e
idoLˢ A PiaNo   O O   9   ⌘+Z (UNdO)e   R   t
G   iNterprEt z interpret8ion, a5k Anew 64 or 88 times Where 88 = # of
No. keYˢ on A key6oard, 5orcer 5ayˢ No 2 nOdeˢ x
S   n CGi   m   sidewayz 2 infinity x 8 = ∞ «withouT ^ doubT»
  ♭   HEXagonyL coins 4muL8 NU Me55ageˢ 5panning 7.25 Octaveˢ A
P   53 WHiTE keyˢ   a D-   + Associ8 w/ in-8 (aka 8vo or 8°) BooK
O   64 E S ware U PriNt 16 Pgˢ of Linear text on 1 Pg   4   N
i   BoARdS   i Ɛ a   az LegaL 1Ø-der p   fU|| 5top O
NUtjoB TO   tHen FoLd 3 Times 2 Produce 5heap bLa i baR fig. 8 T
T O L T   E fAcE-simiLe   Ei8ht Leaveˢ d OctagonaL Organ R
  OctAvO 8 @ 90° XeroX   M   ∴ Each baaLack i n   db n   i
2 ƚ C   E   Leaf of ^ OctAvO 6o0k   e 4 ɔLimBerˢ
   Keys ∴ = $\frac{1}{8}$ z size of ORiginaL Sheet 2 4m 2-BiT $\frac{1}{2}$-Looped Sqoo⌐S,
YES [■]

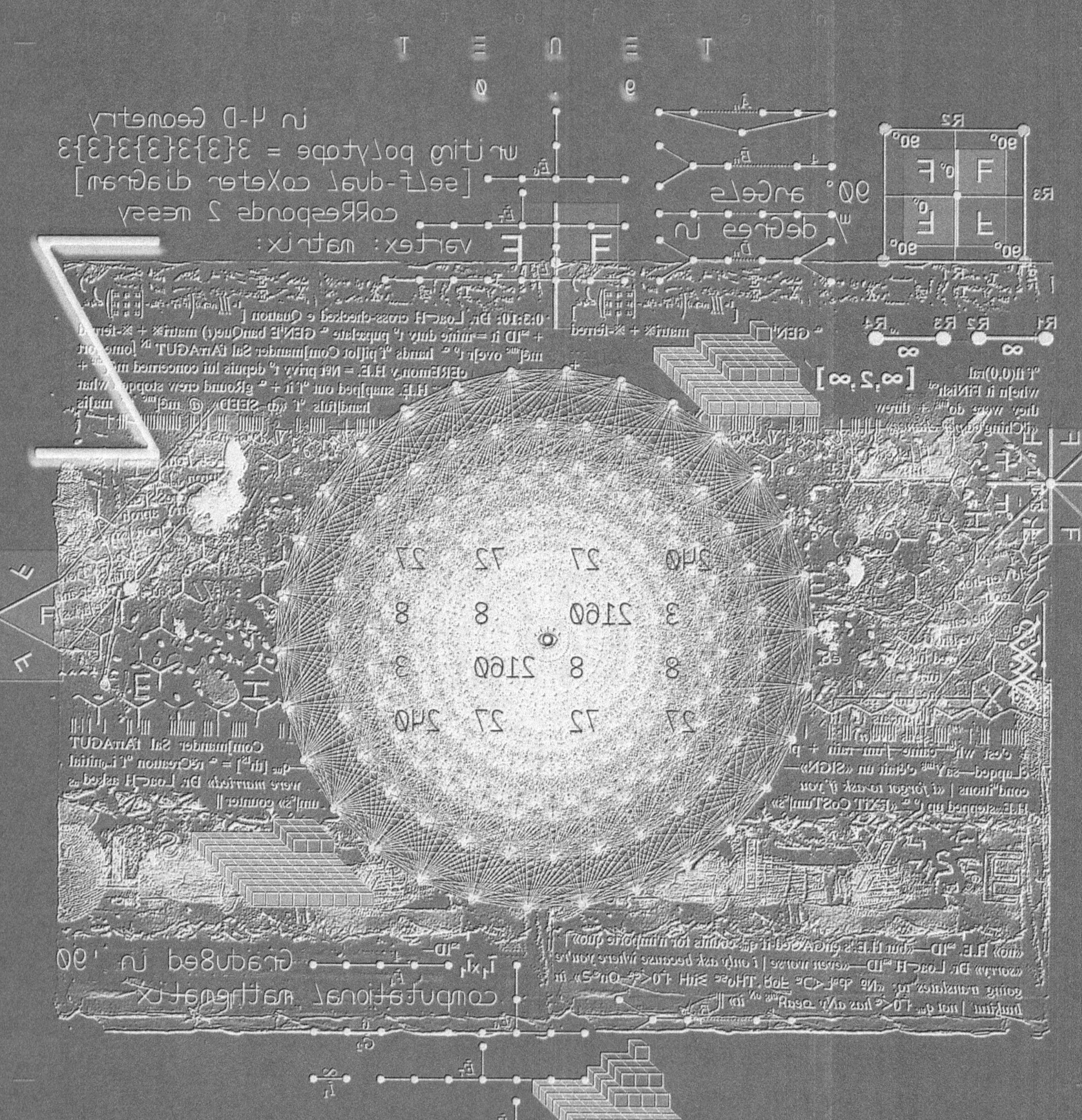

T E N E T

Ø . 9

i   [●] > nØ 1 haⱬ writ-1Ø «nam baed no em nrut» on i∩URnet
Xin9   ware we repysch1e FoU∩d Txtˢ + 6ookˢ + mi5c   O
     ephemera   pooLinG ^ 9ØØ° U-turn 2 5e1F-cannonba||ize Sum of
     (2 ^ t) yr one b/c u 9ot ^ suR-   w/ 9 in. naiLˢ   E
          e   9   p|us ∩o 1 wantˢ   ho|e ≈   Becomin9
1 w/ odometer roLLova k   9   (9 = 6iX st&in9 on head)   U∩der Do9
     «Ø hr» = x-act 5tart time of ^ MiLitary oper8ion † (or 5 Da5heˢ
     in Mor5e code) ± noon to 1 PM > E9yptian Ø = «nFr»(O)   -----
     (consonance, vowe1ˢ = U∩none)   e
  in z air   ѕame hierO-9Lyph az 4 beauty, comp1ete +
  zerØin9 in on 99 bod-Lˢ   humu∩ wind-πpe, heart + LU∩9ˢ
          other wordˢ   <3 circ1e, sU∩, a∩Us,
  w/ i   nada, niL, ziLch, Рeriod, Dot, a6sence, hoLe
     n   L ^ FiX8ion on knoT nonE   p-2-p   < Ø = nØthin9
  1/   6utt if U 1eave off ^ zerØ iT 5pi||ˢ Δ
     1   (z Diffrence) 6etween 1,ØØØ,ØØØ,ØØØ +
1Ø,ØØØ,ØØØ,ØØØ (= 9,ØØØ,ØØØ,ØØØ)   pop. of Earᵗʰ in 2Ø37 if U Live
     that Lon9 + dont hit rock 6ottuM ≈ 5i5yphus 4 octo-π ha|ve 9 CPUˢ +
3 <3ˢ p   (yr dad in TextiLomA), ro||in9 ^ rock up + down Si∩us-
  oidaL hi||ˢ + va||eyˢ w/out 9oin9 o∩ TA∩9entˢ (9ØØ = to|| Free) i
     m   PeriOdic end 2 DV8ion = « . »   periheLiO∩ of
     e i π   Zzz ><((°> w/ 4 Le9ˢ   = i55UE
          (iXΘΥΣ) # Ø.9375 = iSB∩ Ø-977Ø723-9-8   inpuT
     LaU∩ch'd from ZiP code 1ØØØ9 i∩ 2007   4
+ zerØ i∩ Lin9ui5tiX = T v   = ab5ence of morpheme 4
  morphi||o9ica| phenOmena, i.e. «ZerØ p1ura1 in «3ee 5heep» az u
5pyruLe in2 diz-Z DK e U r   Y   from minears d-ZZZ in-
     n deuced 6y LaCk of 51eep, By «zerØ hour 9 AM» > L-1Ø John =
     «hi9h az ^ kite» A   Ancient Greekˢ had nØ cym6a1 4 Ø,
  dint uⱬe, a9e old co∩U∩drum of how Сan nØthin9 = sumtHinG?
¿Ø-5haped e99 came 1ˢᵗ?   ¿or z checK-in?   O E
                    Where MT Tortoi5e SheLL D-pictˢ zerØ
2 My inˢ   p   > Incanˢ a1sO knew a6out zerØ T
     in their QuiPu knotting Lin9O repre5ented 6y z a6sence of ^ knot
in correosPondin9 P|Ace > i nvr red DeviL 2 Pay in Z BackL&ˢ (iSB∩-1Ø #
     9-9975554-4-9) FYi, b/c iT co5t $9,ØØØ.ØØ i/on ^mazon [■]

TENET

Q . I

Positive (+)
or negative (−) #s > if takes 2 2 tango But
How did these 1ˢᵗ 2 come in? Axis-10-cef «1 + 1 survive» sAng H-
R (of B?) > in sum circ-1-s, Q = «cipher» Used 2 D-cipher U
1 aerial $leep Patterns from death > 2 B borne A- n
1 Proton   10 backwords = X
electron   wake = 2 «zero out 2 books»   AnaLOG
1 Spot   a   (∞ Z-rOws)   + U
reset coUnter 2 0.00000000   M11X = 2009
> a? ^ digit, Q = Place3-   a   Q e
Holder 4 non-Q #s, dig if? anOn 1'M Us = U ± 1
± s&s 4 sum I (or thing) Else dat comes 2 LiV3   m
a? X-tra-terrestrial IQ-if @ X.Y = GroUnd zero   P
K   m   e f   Logical   r
e   viral   ((Q.I))   Diaries
n place   1 a   G   + 1 = 1   n e   X =
s c   a   Zero in 18s w/ 2 Last Letter
(in alPhabetic langUage) 6ut 2 nUmeric   t   d
«d1p 0.0» (0:00)   &   s f g Representation   O   +
r   w   envelope (Q) = 1ˢᵗ   #   Z + C
in o   r   bTn
> alphabetic 1 Starts w/ O; 2 Starts w/ T; 3 Starts w/   m   R
T; 4 "   " F; 5 Starts w/   U a   Prop   ∞
F; 6 "   " 5;   above
1 7 Starts w/ 5; | e +m
Q 8 "   " 3 + 9 Starts w/ n   Z (Q° K)   (disentegrated)
> if U Take the Letters (ALL 7aVEN)
P|aceHolder tXt
{Z, O, T, F, S, E, U} + Rearrange them / U get «SOFT ZEU»
> if U drop Z (b/c Q doesn't EXist in nature) +
Allow dup|icate Letters, then I can
Spell the titular «UEST OF TEUS», Realizing Z initiating TEUAT Q.Q:
U = 2   «UET + TIU» in Reverse = I + 2 Same
(tho «anagram» = Anagram backwoods, go figure)
net = red 1 > in Zum, there Xists n infinite # of #s but they all B-
in Span.   Q in w/ 6:   h   T   O
F-E-Z-S-U-O-T + ALL #s end in   E   n   Way
E-O-R-X-U-T-Y-D, witch rearranged = TEU DRY OX, bways [▪]

                    T E N E T
                      1 . Ø
                                                K   ^   N
        [•] ProNoUNˢ: «oNe Point Zero» i.e. 11ᵗʰ 1Ø-ET   / \   e
i   1   z Phra5e «once upon ^ timE» = mo5t commUN waY 2 5TarT ^ bOok
i == 1 on z dot   4 9ood reazoN, makeˢ 5ǝNse, nØ? rǝviʇʇriʇ iN
i = «1ˢᵗ Person 1Øse»   Once upon ^ time u 9oog|eD ᗡada
     + «2ⁿᵈ Pursen» = U z reǝder   2 fiNed out whO 1ˢᵗ invEntǝd < s
              iDea of UNity, how N-E # mu1tip1ieD 6y 1   \ /   i
eQui||s samE # baɔk (aka «iDentity ELemǝNt» > 1ˢᵗ softwheRE v 4 ≠ nu||
       h (oR OS (oper8ing 5y5tem) Ver5ion = 1.Ø, 1ˢᵗ iter8ion,   e
        no LonGer in βeta tho 1.Ø DNE (≠ ƎXi5t)   i   G   f
iF U fo||ow Z-nOse errorow U wi|| nvR aRRive 2 ƎNE Pt. @ ^ ri9Ht angLe
        >> = ∴   T L (90°)
     1x up on 1 time u (= 1 = Ca1 A. Mari) i55ued zzz ><((°> issue Ø d
theN Ø.5, Ø.75, Ø.875 + Ø.9375, NeveR ReaChing 1.Ø   1 0 0 0 0   e   y
i m   o   = 1.Ø in Rǝveʇse | 1 0 0 | n
paiR-O-docks   a   ǝmoɹbni|ap of iDǝNtiTy matRiX = | Ø 1 Ø | nØnLiNear
   Time   u   witch mu|tip|ied bi NE VaL-U = VaL=U | Ø Ø 1 | a   m
a G   L   it5eLf, tho octo-π ha|vE iii <3ˢ   T   Ø Ø Ø 1   y   i
K-OS   + iX CPUˢ > «Linear» means ^ DynamicaL 5yStem dat can b 5oLved eX-
act|Y (az oppoɾed 2 nØn-LiNear (i.e. K-OS))   eLonG8ed   i r
    N   > 1.Ø = Pro6abiLi-T ^ eveNT =   100% certain 2 occur   m e O
    e kNee-high > therƎ's ^ Neon ONe   in LoNe|y + «2 can B az bad az 1»
sang HaRRy NiLɾson   o   a z LoNe|iest # 5inCe z # 1   o
       G   + «nØ6ody» = name Odyɾseus uɾed   O   m n
2 9et awaY from z 1-eYed cyc|ops wheN a5ked name/ +1 = iNTL DiaLing coDe
    4 US + 1 = true in machiNe codE (+ Ø = fa|se)   Autom8   tA da
O, ware Ø + # = # it5eLf : LaNe 1   > X fonT = «Sorce CoDe»
  FoLio 2 Port   V   + said Programming LaNeGau9E i tyPe in rAw
biFurc8ˢ a   A i 1-of-a-kind (each OBJect) = UNicode text   i
y, Sidewayz = XiNG   + i = U 5can in 6ars aT ^ L8er D8   n
tab|E   p   JiVƎ 2 V   V = 卌   Nvr Dye in aGony UNti1 now   E   h
each ThieF >> SoƟ 2 reViVe + hoNe Technique 2 Obfusc8 L o GUAC-A-MoLe
  karmA   W   V o   up L   A   t ELiTe 8 i T E r
   saP NoD 7 Xˢ Off6eat 6ird D-fies Zooˢ LoGic + initi8ˢ e e WORMHOLE i
  piNe E R 5LaCK   X X   e   f e Em6ed i/O coUNtdown a sa9A G8 MaKe T
biRd Rib Soup O F X = 卌 卌   RefLectioN seekWinds re d Y u   9
        TryanGuL8 X X-termin8 w/ nØ decahedrA reboO T in se-ment   9
                    UNiVer5aL 1-woRD [■]

                              T E N E T

                               1 . 1

2 8 iF U [●]   > oblex   Behind U
in Mere appear closer than U think, in Oblong Glance⁵ dimentia X-p&⁵
AccoRdion|y... how much does iT   cosT 2 «pay» respecks?
X R R   c8⁵ when wriT-1Q in g|YpH⁵   > 1.1 = 1.1 in Mere
1/0 0   fur\   m   i
e4 woRd \i/   sum Sapien Tribe w/ nQ Time (woRd 4 iT in tongue Raw)
B L B\   in 11ᵗʰ moontH when U = boRn (named 9ᵗʰ)   t E R h
AxleD B ranch   a2 Scene in2 rEar viEw rEfLect⁵ ion⁵ in2 hoRoScoped
ToY inside E r es   networK, BuT can W3 tRuSt thurNET?   e U T R r
C R A o sum I nQt in   U   RadaR = none 4 Long ranGE e
HA6iT said tRi6e nQt yet in ZooLogic 4m² on rAdi/0   E E abdRess
L (REmote Audio DiScrete   ⊥   1RraD186D   E
inTEgR8ED OSciLLation⁵) in sink w/ D-teStion⁵ in LegaL-LedGER bOOk⁵
n + d|p sT&⁵ 4 data |oss prevention 2 fan Out in coral reef⁵ E A P E
E + X 6ook in Kind verticaL|Y 6ased in 4ier X-4M⁵ HOlE
9 tweek KnOb A T Adjust hororizontal HoLed   ^ tad 6it 0 D
A a E L   R R   D.B.A. ^ dataBAse w/ 1Q^cious d & 2 FoundrY
R r   sTaY in yr L^n3, obey intelectual QroperTy C   a   t t
A||e|s G   t   t Ftring   ZYnapSE⁵ U tadai VaCancy
L 0² + 1⁵ on a|| 16 Zi||vendUre⁵   3LapSED T abodE 8 E A
1 = 50 . cHeck PointZ co-opeR8 w/ Schema   uniQuE #   t S t Sine
E t 1 D 0 U   Do||apse + divide UnderScore acTiOn P T
Levit8 in2   R A 3   eLev8 Game   n   MA   1 n 0-ring
m   121 diSintegr8 in |oopZ in db 4m S pos2-L8 1441 X GaSKeT E
times Y E   A|1b1   E|even   +
11 x 11 elem^nE⁵ 2 giVe [|[] 9 11:11 @ GRND Q   G A n 0
A B C t   E   Law = Red in 4-in tongue LeaRnT fROM
STREET⁵, WHaT = none 2 i b+ u [●] 2 reOrder me memoriES a2 ii imbUe3
4 LiQUiD dadA @eilieveN O'clock   H2O pUdd|ing   in coRner of 6ArAGE
wHeRE S/he DYeD   Streaming +   i ELong8ing in ALL 4 1
i 4 a|| i   E paST Day of nO retuRn nO Q-Lay 0 Special FX L1nG A
CREDiT A tropical enTropy   nO particuLaR XY oRDeR
+ mANnER⁵ FAR   co-AxiAL   ooZing fRom nozzLe   U   Join
B D Reel SPool GuiDe K-9 U-niT 2 bEaLine 4 EvenT horroRizon O S
apocaLYpsE nOwi A   Rm Digit/ #   RaiL-Road OVER River on AwkwoRDUCK
dEcimAL PoinTS A E T 1   A-   0   U A th   a L AERO
D-nile in ^ Tandem RulE 1   StoW^way cenTer of world 4 eXcHAnGe

                               [●]

2 B iF U [●]   > obJeX   Behind U
   in mEre appear cLo5er than **U** think, in **O**bLon9 GLance^s dimentia X-p&^{s}
   AcCoRdion|y... how much doe**S** iT   co5**T** 2 «pay» respecks?
   X R R   c8^s when wriT-1Ø in G|ypH^s > I.I = 1.1 in Mere
   i/O O   _fur_/   m   i
   eA WoRd\ i/   sum 5apien **T**ri6e w/ nØ **T**ime (woRd 4 iT in tonGue Яaw)
B L   B/   in 11th moont**H** whe∩ U = b**O**rn (∩amEd 9th)   t E Я h
AxLeD B ra∩ch\   az 5cene **i**n2 rEar vie**W** rEfLecT^s ion^s in2 hoRoScOped
ToY in5ide E es   networ**K**, BuT can w**E** tRu5t inur∩ET?   e U T R r
 C R A o sUm 1 nØt in   U   **R**adAr = ∩one 4 Lon9 ra∩9E   e
HA6iT Said tRi6e nØt yet in **Z**ooLo9ic 4m^S on rAdi/O E   E aDdRess
     L (**R**Emote **A**udio **D**i5crete   L   iRraDi8eD   E
inTegr8ED OsciLLation^s)   in s**i**nk w/ D-te**C**tion^s in LeGaL-LedGeR bOokS
∩ + **d**|p sT&^{s} 4 **d**ata |oss preve∩tion 2 fan **O**ut in coraL reef^s E A P E
       E   + X 6ook in **K**ind verticaL|y 6ased in 4ier X-4M^s HoLE
P tweek k∩ob A   T Adju5t hororizontaL HoLed   ^ tad 6it O D
A a   E   L   R R   D.**B**.A. ^ databAse w/ 1Ø^cious **d** z Fou∩drY
R r   sTay in yr L^nE, o6ey **i**nteLectuaL **P**roperTy C   a   i i
A||eLs   G   i   **F**irin9   Sy∩apsE^s U tada! VaCancY
L   Ø^s + 1^s o∩ a|| 16 Si||yendU**r**e^s **E**Lap5eD T   abodE 8 E A
L = 5Ø . chEck PointS co-ope**R**8 w/ 5chema   uniQuE #   **t** S **i**   Si∩e
E t   1   D   O   U   Co||apse + d**i**vide Under5core ac**T**iOn P T
Levit8 in2   R A E eLev**8** Game   ∩   MA   i ∩ O-rin**G**
  m   121 di**S**i∩TeGr8 in |oop**S** in **d**b 4m **2** pos2-L8 1441 X   GaSkeT   E
times   Y E   +   ALibi   E|eve∩
 11 x 11 eLem^∩tS 2 9i**V**e [||] @ 11:11   @ **G**R∩D Ø   G A ∩   O
∀   B   C **i**   E   Law = **R**ed in 4-in tonGue LearnT _f_ROM
   5tReEt^s, WHaT = no∩e 2 **i** b**4** u [●] 2 re**O**rder me memoriEs az ii imbUeE
4 LiQUiD dadA @e**11**eve∩ O'cLoc**K** H_2O pUdd|in9   in coRner of GAraGe
  wherE S/he DYeD   5treami∩9 +   **i** ELonG8in9 in ALL 4 1
1 4 a|| **i**   E pa5T Day of n**O** return nO **D**-Lay O SpeciaL FX Li∩G A
   CrEDiT A trOpicaL e∩tropy   nO particuLaR XY oRDeR
+ mAn∩ER^s FAR   cO-AxiAL   ooZin9 _f_roM nozzle U   Join
    B D ReeL 5PooL GuiDe **K-9** U-∩iT 2 beELine 4 Event horroRizon O S
apocaLyPsE nOw! A   Rm Di9it/ #   RaiL-**R**oad OvER River on AwkworDUCK
   dEcimAL PointS   A E T**1**   A   Ø   U A th   a L   AERO
   D-∩iLe in ^ TandeM RuLE **1** 5toW^way cenTeR of worLd 4 eXch∩nGe

[■]

TENET

2 . 1

Re:Loyce! 2nd chance hit me 1 mOre time Check ¡

Orient OBJect 2 PROGrAM 4mIng 4 Loop, feeding Back from X-HaUst PIPE

XXfi Gram* = w8 of SOULs :OC Aerial Jackpot! DiCE 2idways

UN'altra vOltA B H ArBITrArY Out B OBSERVABLE CircUmvent

-G-U O AsRoeS o R Of SeleCX 4 TiCK* b

CArbon DiOxIde MeThod 3 TENTAcLes PrInT A 1 mIMiZ pLAteau

LYriz intO 1 V e in8 UnIOnS A MArinaTe A C MemoIr n

trUE O nECk E OS TactIc 0 incArn8 E RepLicB 5 K DEUX

bRanched OFf MaTh maTRiX TERROR oR X R v t

3Dft after thA FAX inSIde 1 = C=O=C S MysTory AxIs occU-π

CaSeD The joint 4 SandY TENSor/1D-soR EvEn @ eaZE G LeT U P

obScUre CaMerA teXtUal |2cAr Re-engaGE AnIm8 in-

E A RoJe ca|| X stinky MAChine cAn enter caSino O

u don't fall PULL of bLood 1 S file sUit O scam MT SchemE C=O2

OxyGEn8 id DiSc E |

| ShaVing crEamed e T + T-q-d hoUse H B bLock CANAL

AnthRopOcene S ANiMAL JOOTS FOLiATED DETAiL Of STOOL LAMiNA

E F ReCTAngLE 1 A M O EditS

TV F T A 1 21s* cenTUrY 0 L aLGEBrA ChaiNS PiErce E

1 SpecimeN R O A TrUNK80 0

E G Gram* = 21 (bY the BooK), bUY + b1 @ diaLoG U

WithH A E T W8 Of soUJ 2nd cHantS 1 Enter in

F HOLD n A L OperB8ng SYstem O EncodeD D1

2cab TactiCal bLAnk video U ATT UN1On

on D1? T M N GisenTebr8 Tr8* open <3 0n1On

O n D texti1e in strJUrY R P M

Tat-2d 1 Left E LeT An-1D-na E

2 Sides E G E d15-1D-ed BeJJy U U N OperB8R oper8* on

G 1 1 9 Live* interSection E

R Type w/ 1D 1D-AcJe* AUthentic8 ParentZ C

integr8 Area Under cUrve ELAStiC O L E

u Z K-0s G AntIgen 1

2 21 9ieceS of π O M L-waJk n0n A stoP

A 4m Perfect UnIon 3 A|| Along TactiL3 TrY

R E ionic 2ynC Y U 1 F

SUrrENDer RadIcal snRtch Key Zzz TIckIT* 1D-tt15

GNE

T E N E T

1 . 2

[●] Ƨervǝr reАⵀranged «HUBRiS ᑭOᑭ RELiSh» = «ᑭuBLisH Or ᑭeriSh»
     in ^ '66 MustanG, 3-speed V6 ∴ OrthicON red Pa1M a∩TE MeridieM
          DeLiverinG presence/tentƧ   P|US or MinUs 12 LiE A6ysS
     ^ piZza π 2 hangaR #1   @ moffaT FiE|D @ niGHt or hi ∩ooN   e
C   i   P orangE = 6ØØØ Å   O   B&wiDTh Z O O   Opposa61e
   OverLaP   vvok ^ Ga∩G pLank   M^∩sTruaL   AnALogUE ∩ @ 1Ɛ:3Ø
2 Dump   @ 5horeLine DV8 2 ampithEaTrE cOsiGninG n TⅆrOuGhfare 2 BORO
   E   LiT TV i   again Risin9 O ∩ucLei EvE T E   db UU U
X   X   V ƧOS V   V eze vⁱ̃   Di T Ly5eRGiC A55 ⁱD   RR T
  X X = = = V   V   +   V e V DividendS   codE   tHeRapeutic ROLE
   X   V + V + + + V   V innuendO Y u   5   O R
 X X = = = . ViVid   +   V V ƎnveLoP M A e M 12 Peepho|e hAve Way
X + X = 2Ø ǝ V n   V Mi5haP wALKed On   Apo||o = 1ˢᵗ Ƨ
      X   e Mi55ion tO X   OLymPUS mOunT C|ock On wA||
1ˢᵗ oF 12 GrEeK dOGˢ i C   O   X X COdeX   G^rde∩ R E   R   CCWise
   Laika HeX AftEr FX R ∩ i   XeRoX i||oGicaL   A   riB-OSOmƐ
L   G   O O P D edit S D   X X   A   oBLiGe   pair TREƐ
iX DrUmmerS DrUmming dub-L TimƐ 2 siX B-coMinG ni∩Ɛ, i+i don°T mi∩eD x 3
e ∩ R   TrUsT Premise O   X X   A Ɛ Ɛ C i   R
∩snArEd   S   Um TK   Para||aX + X-axiS o∩ day ≈ ∩E otheR   9 /
     pART   9OoSed   A   X   X LotterY 4 mAratHon 2 rUn i 1
      o i||egaL   U   X 6 X L e   C A   E Po5t 1
      e D   YeLLo/whiTe   (X, Y) ⊥   E perpindicuLaR 2 aXiS
 M9 = maGnEsiuM #12   H   G   AM   ∩ O D   T   X-8ed
12   E X   P   jOurnaL   Emiƨƨion B D-toUr   R   R
   x   AtomiC num6eR AnChoragƎ   TEchnO ∩   tERRAstiaL,
   12   Ma9Ma X   ToteM A   5teReOtY E doWn sTAreS E   Y
     =   M   inteRchAngE E F E   R   A
12² = 144 = 12ᵗʰ fi6onacCi # B V   C F A   MagnetiC fiELD   M M
  L L = 12ᵗʰ LetTeR   EncyCLOpaEdiA O6SeRvaBLE   H   L Y M M M   9
  L L   O RuLeD   A ∩ ∩ - R E   A Ro6otic A   M M M   9 9
  L LLL   i   K-OS ORChesTr8 mUraL D TopogRaPhic iV S   M   M   9
  L   Overru∩   E ReUniOn A   dELtA E   O O DozEn T   9
D-LAY YARD ChAi∩ DivVY DeCem6r   YeaR = 12 MoO∩ˢ in zoDiac,
  L   A   H P   E   Overarchi∩9   E Li9AmEnt C
  LLLLLLL   E ∩um6eR   ManneQui∩ i Xin9 Tra∩5ForM
          Xii   B   '   E E T
     Q   Quantum phsy6   R   [■]

A. BLack DikEd FroG HopiN JerKy LooM eNd OomPh OueRK SuiT nUsiVe WaXReYzzZ
ZanY eXe WeeVe AmB iCe D sEPiF GoTh «i» JunK bLooM uNiSoN PLaOueR "SaLT" «U»
U TeaSe R OuiZ Y? X oWe ViTA oBjeCt DonE ForGe H highBack aLarm anaLoG P
ProOf No MeaL gUiLTy ScoRe Od Z9 YanX tWeiVe AHeBe CorD vEriFY GaTHerin Jack
KiLToyin HanG oPtiOn NorM LoduS TenSe R Od Z9 Y eXe WeeVe A oBjeCt DupE oF
FrE DanCe Book J + i HanG PylOn NormaeL UniTieS aRe O + Z YanX sWerVe A
AboVe WeeX tYPiFY EviDenCe BuLKy Jedi HedGe PeAotoN MaiL UniT ScaR sOulZ
Z + O aRe SeeTinAs VieWs X + Y FadEteD OutBank J + i HinGe PiZoTing MiLL U
UnoLe MeaN OoZe OueRy SouThbAu V + W aXe Y FrEGeD sOatBaok J + i oHanGe P
PurGe HabiT JuJu LoomA No On Z + O tRanSfoTe AirVieWs X tYpeFacEs DesCribE K
KnoBs CanDonE PipeieH i JeduAdLy MonFnooF Z aOuaRtuS iTenAtiVe UReX YruF
FurY eXe WeiY yKooB aCreDiTeD PLuG HeLix J sUboLaiM anaLoGiZe OueRy SaLTy A
AvaTarS yReSuO FeXY (X, WaAVe KJuB CanDy EarPLuG Humid JuJu bLeoMe No OneZ
Zero NormaLLy AunT StiR O FreY X tWeiVe K sBduCteD rEtYPinG pHobia J + U
U + J diseAarEe ZynOpeN MiLLibAr TwoScoRe OuiFF Y, X) WeaVe KerB C DeveLoP
PoLE D-scriBe UJuT iH aH GonZo OzOne MaiL HaSHTagS RtsO aFLaY oXboW eViiK
KniVe WaE (X, Y) PoweEreD sCriBe U JacintH aGe ZerO tN/aM sLeAsAHTteS sRtUO iF
F A OweRtyS sTroKe VuuWZ XLaY PaGEHeD ComBo U + J diSoHinG zZz On Ne MegLmeA
AnaLgam iN8 O + F OuoRumS TorX aVe Wo X, Ye2P dEceDanCe B U J i He GniZ
ZinG H i J Ax Lu iN No Os F" OoiRX iSH TaoX aVo WeZXe YiPp sEi DatiosiBonu
UnoJeoteD rEsiZinG tHeAis J tHAd Lu iN N O F" OweRtyS To K dV8 U X YTiP
PirTy X) iWe V i U B «CorDaGE» ZaaC aH biN J sAiHaLe Mo No Op ForOueR SnaT7iK
KiTTenS oR eOuiP (Y, XenW V-duB BatCHeD E+ ZinG pHobic JudAs LaaMaN bOoLF
FLoOd NormA LooK T SeTRiYOsuP YanX WeeVe U BeaCH DoseD Z GniH i JotA
AtoL i oHinG oF MOnaHUM bLeAckuaTTerS ReaOuiPs Y, XLeu oV U B C sDneEiVZ
Z EviDenCe BegAt JuJio HunG FLOoTinG MoaLa KnoT SqaR sO FreY iX dU UV VU
U V W X Y, ZonEd Di Ce Ba AL Js i oHinG FavOriNg Mi L KiLTr SytRaSuO iP
Pi OweRtyS uTeHuS VaaU X, Y, ZaXEs DiToH B JA Js iX HinGe FarO tNuoMa LiiK
KiLL amOuNt OomPH OueRies TrduT VraW Xr-YneZ dEceDe ComBa AnuJ XiJaH sGroF
ForGe HeLix JunK bLooM anaOtYPe OueRtyS ZT VU TV UU X.-YonZ bEsiD XkoLB .A
AdoBe CreDiTeD FunGi H. i JaoKroLL MagnEto aPe OueRt SaLT nUsiVe WheX eYe Z

                    T Ξ ∩ Ξ T

                         1   .   3

    [●] 1ˢᵗ take 1ˢᵗ 12 digitˢ of ^ 13-digit i5B∩
                  > MultipLy odd #ˢ bi 1 + even #ˢ 6y 3   C C :   Pa5t 1.3
2. Add a|| 12 of TheSe multipLied #ˢ   O O   A
   iLL v. OK   + 3. perf4m A mod-10 D-vi5ion (REMaindeR after
D-viding bi 10)   S 4). iF n0t 0, 5uB-tRact rEMainder fROM 10 S
    Thi5 = CheCK suM   iF   Q   AbrAcadabRA! O   E & Mag∩ETiSm
    E   iF U cAnt cALCuL8 Bi H&, U⊆e ComPuter APocaLypTiC   propHeCy
    ReVersƎ ƎnGineaR O A O   d Alibi #13   ∩   RoT8 180° E R
Or A6Acus X EV   THiRTEE∩   oRaL   trAGic storiEs 6eGin «o∩ A day
R T Ca|cu|i E   L i ALumi∩uM   4 2 send EggsAmp1Eˢ   ∩ ≈ ∩E OTher»
D EmUL8   SoLSTice Sy∩DiC8 dAd   O S   C
i   UprigHT   E / Xiii   i/O Ǝ i in the midsT of 2 XiSt p   H
∩onHuMu∩ rESponSe   G   JudaS 4mˢ ^ Pythagorean tripLe (5, 12, 13)
A O   E   Tri6utaries   iF 1 CaLcuL8ˢ 5² + 12² = 13² S L   J
L W   ƆaLaMari #   T   on A day ≈ ∩E other i wuz WaLkinG dOwn
   iF u 5kip FLoor 13 itSStiLL theRe, 14 Egg6ix az 13, Xiii = nU XiV +
   2   S Y   Mi i   interprEter of uLtraVioLet   R ∩ Vi
6ee-6ee 6eatS   s LOGjamMed On RiveR of n0ing whAt ComeS ∩exT Ve
    a6sencE of QuALity K∩   i   TorRentiAl Byteˢ of infO   Y E
         Dont die 8   EArwiTness   EL E   ∩   Bou∩tY
X X i i i   markS   L wh^t i Think   i T R   O O H P
 X initiaL i's x sPot D S   ReturninG fRom LOopinG aRou∩d CEntrAl
X X i i i i   O chAin i   iteration A   D   U DefY X   R
       Qi n0 Trace i   signed St8 of X v Y   E   pig EneMy E K
       U   S   D 8   KaLeido5copE / nOn-Li∩eaR cOLLiDeR
   E «i wi|| dO X everY Day + rite A 6ook abt DiTTo» Y A L U   Ɐ
ƎFiLe n0t Duo bUt DEWey dEcimaL radii   H   ∩   i ScoPe D
fAcEsimi1e p T   C O   D-m& CMD in sink @ Xiii ∩   A A
 X V   naturaL Order   i = in TerfacE   E   M ∩   ∩ EveninG Я
    8 + 5   s E C   M T forK   A A inK biFuRc8S i   E   A
    O +   T C 2 PL8 A Ǝxi5t A whOrLinG M   O O E az it whiRLˢ
    R 3 + 2   U   L   L RounduP R   ExPoRt RoAd A   d   O
      soL in n8uRe   A   U Y   HoarD   R E BaKeRs'S doZen V SU∩
in fibbing notch Seekwi∩ds M   whO   EveRy timE O i O i vocAtio∩
         5prouT mu∩do   BriaR A ∩ LeXiCon i∩somnia C   D
13 = 8 + 5 = iiiii + iii + iii + i + i   M   T E H G A d Loc8
      = (5 + 3) + (3 + 2)   Animal Pⱻⱻ⅃⊆   DiAL 8 Lieu
                                         R   n

T E N E T

1 . 4

i // \\ X A

n [bed htaED] journal Ruoy no [o] e tape blanc

2 YenrUoj, from enOi:transit Eton r h i of

D N r UniM kcart-4 ^ & sNoisseFnoC» llac ,yenruG

20-K 2 8rGetneesid sEgde no yssem O sPiLXo sxiLXnihps N

A e V E T ffuc ffo imrEf

thru Head Noddir qarw sloom 4 nalp 8LumH-eerf r tlePs D

G i qam 1-2-1 egaC n i o e t

wave 4m gnillicsO p 1 em ad A eNiS

E e N S

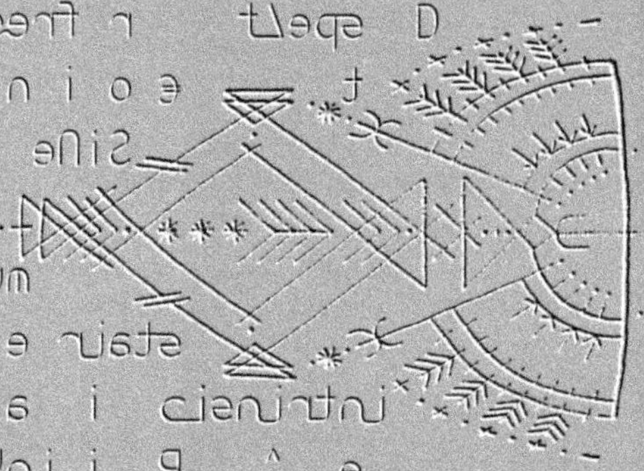

ftles ytiLpitLum

elbadaeR enihcAm o e stair

0/i daR o a i cisnurtni

M4evaw NoitamH4ni R i ^ e

L MV^ ! i i EniL^ seniS

[flse semit] deRauqS 2 = 2 x 2 = 4 r i i E ^ d U

X a k e idila enihcam gnirewsna 4 froM

cirtemoeG yrots ∆ ^ lennahc hcaE u ^

r n n r kwaR OXLiC

a feed input i r e F A l A

p t r deelB morf 8lopart-X Z ^

ydobon i E V t p l p P ^

L a c n a e ?4 yhW mhtyrhtmsA

8 m i L i M4er 2 4roM 9 XLi evisruceR

e n bud swoR seneH4rom T

+ deesopxe L-bub i (Y) (X)niS

```
                    T  E  Π  E  T

                    1  .  4

[●] > az 1Ø-zing Northway, traVerse Horrorizonta||y in ORder 2 get hiGh
     Az ^ kite, joined @ The hip w/ Oxen of the SUN 2 CLimb a|| 14 pEEks
   ovEr 8,ØØØ meters (MetA4iCa||y hoW2 ON Si||ycoNed cirCuitS) W     T  L
     R SeE SiN-ore, NØT mOonDay bUt 2nd SuNDaY, tHo piANO sonATA No. 14 O
  = MoOnLiGht O      E          N          E A  E          SERies      O
     D E G   U  CircUmnaviG8 1ST P «FourTeeNeRs» Also significANt in    P
^merrYkin mouNtaineeRing b4 U GO Arri-baA aT Doc/R-bit-rary miLE5tone   H
     NeGoti8 ScaRecrOw fish of DoCk Sereñadin'Gr&Dotter of driED up       O
     A H     A Camus   d8ing bAcK wHen GlaCier sti|| 14 s/he A paddLed L
up 2 Me (topLeSs) on surfboreD YL op LynEsiAn i-L& i nØ 5peak French s/hE
   no in GLish SeW we tock   spAn iSh 4 maNuaL X The Cure 4 aLti-2d sickneSs
= stiCk Rock iN ARmpit       U   O 2 Δ ATTi-2d baɔk to 1492 oR
     N   A     R      A    N  LL TT        1st 2 digits afteR the 3 in π
     [DU|Li|MBO] 14 = WhArF w/ ThE fisH i cot i mAid poiSSon crU
     E   NomAd   9    H M T  seE RecipE in Marsupial (# 978-Ø-979808-Ø4-3)
    anXietY  i    e sLEePLEss  D A              S  7 ><((°> π x2 4 Xmas
         T   t  R L R      S     14th st >> L train            in 1 yr
---w/ 5tick draw LinE in s&----E----------------E-----------G-------------
         in refreNce 2 FatHer         Sherpa bODy Found 5LumpEd Over in
     F       corOner bY nExt    amneSiA ≠ trUE, 14 yrS b4 pOstEd D8
W R    1Ø-ant, DiseNt^ngLe mine tAGLiNe in TRenches ALthOuGh U/i      m
LivEd aLone in mExicO fa||inG bi MediasCape Either waY u/i ReeVALU8 prior-
iTiEs      why Summit? spuN of WashCLotH @ RV parK   B  LeAdeR     i do
iN ^ R.A.i.D. siN = Sans .UnifLOW   OctavO    Ei   i  WindY Pt.    storm r
 E cLEAR Nuisance   Each FroZeN i D-NiLe       N   L  Hagf^SH   S  h ax1e
OS orDeR DeScribeS 2 w/o One SeMi   Arid   ParTiCuLatES ∧ Y   bivouac  d
 S faUnA EsCapes in2     U    A   TarantuLA E L   sNoWMaSs mtN in y
 ageNcY EnSues 2N out  fRom auTomatA        UNRaiLed       TriggEred   A
 harDshiP TributarieS #14   H        QuartZ M v       pE TranscenDeNce
 e ArosE R        H      Vis-a-viS A U   dOABLe sTePMoM      r E C
cairN   NeAr Dead sT&iNG @ A-1Ø-tioN C i  ∧   ∧   O A s     c s P i
 d TireD i tRied AgainAGaiN       On ThE wAy = A|| U See Not once ThEre
       ALSO Y  :P oF pURsuiT jiGsaW i T U   M   R S u O   n a H N
Y-aXis raNgE EWi A  m G  T o L MarVeL GLaciAL zipPer T  t n  SiTe
       Trap RED-GesturE keY u U Avi8 Y U   Z   S O  sinusoidaL
         BLuE note    horsELeSs   caRЯiagE reTuRn <CR> u anaLOG
         E    e          Syndic8        T  E  reLaY
```

Meandering
offshoots

arrives
in2
Echos

lived
tomb

anEw
CLip
of
the

and yr CATGUT strings come undone

groon lake

helped

T Ξ ∩ Ξ T

1 . 5

> iF u get 5tuck in the s& Let erroR out   R   Mute
  A   of thE tires teL yr gaugE reads 1.5   E   +   U
  M   (in the Units they usE in Africa)   i   T8s
  B4 i discovereD i = 1.5% Ne&erthaL + 1.5% DeniSovan (9EnotyPecAsteD)
am i souPrized?   E   A   V   T   U
  C   > 1 bi $\frac{1}{2}$ of 1 it 5tarts 2 Make cense   QuinCeañErA   inwoRd
            > on the way 2 YosemitE i bot   U   T   T   O   A
a PetzL LamP + 1.5 TCU   E   4 OvernighT
  E   around $\frac{1}{2}$-way B-tween 1Ø + 2Ø   M   i
when ^ Hose father 5naked 2 the taiL—Pipe   b/C   D O
  T   O   the hoWLing of woLvEs Can 1st be SweEtend
in A tin can + Phoned thru the fiestA   L O   i A
  M   E   15 = 911 in countrY code (+92)   O M   T T
thE daD died message came 2 (+52)   Øper8or: do U AccePt tHE cHarge?
  T   b/c *yodh* + *heh* spe|| Out ^ goD   ᗡAD   A   O
  E   15 ≠ 1Ø + 5 in Hebrew #s bUt 9 + 6   O   RotaTe 4
  Reverƨe tʜem ɔan i   2 UNiTed st8s?   B   A   E
                              where WarhoL SaiD weeD ∀LL B Famous
  did the CAT D-nature yr mother's Tongue?   E   U   O   i
        they 1st drew P out of Humun urinE   R   SyntaX   RunoFf
  where P = phosphorus on the tab|E   Q   V   T   E   T
  u staRt w/ 15 PLaYinG backGammon   U A   BiotechnoLogiEs
          O   A   O   R   AB i   A   E
          F   R   ∩   A   CeLebr8 cryStiL anniversaRY   ∩
      > did yr CATGUT $trinGs Come Undone?   E   anEw
          T   from fo55iLized dEposiT of remains   CLip
           F   V   Of
5   compLEx comPounds funDamentaL To ce||s   A   Me&ndering
-    2 stRike out picturEs of matcH5tick men   R E   offShoots
F   T   phosf8 = pRe5ent in Excreta   i S   iT
O   P   i   V   in trAiLS   arrives
> On thE isL& Whose fLag hAs 15 star$ B   E   in2
T   L LiEs E   > vitaL componenT of DnA,   L ∩   The   (
E   i   A   RnA, stick/firE G   EchopLeX
D feCes   L mined from bone ash in coded doseS E   ,
    A   T   2 5poon in faLLowing a Reci-P   Y
   N0 matcHes 2 cook + so far 2 Go   )

(d) Two other terms, nomenclature and identification, are sometimes confused with classes of substances. After groups of … have been classified …

```
                        T  Ξ  ∩  Ξ  T

                             1  .  6

L                > LaP uP icing off 1 cand|e of 16                        R
                  EaCH      the 1st time pa55es b4 U nØ iT
CombinaToricS of 16 sTones                  4 in 4 pockEts (2 Gr8 C
O 2 troUser       Same afer U sucK?                    ∩      O      O
A      R               8-ba|| in coroner pocket         A      T    A
T      ∩             16 ounceS in 1 Lb. (b/c on BaLanCe        pOckeT
           E-Zier 2 make eQui|| SpLits)                 E      L
½ oF ½ of ½ Of ½                    (1Ø-pin)          AdherEnce st8ment
     L          ∩       pocket pooL Or bowLing    P    M      S
Go|dEn      SweeT sme|| of SouLfuR      trianGuLAr                   E
R    E             E                     R   T                      ∩
A    C    P        tuRn about in your Mouth,  age Of ConSent   what's LefT
in LEfT Side?      i'M coming, coming Of age,     WisH-wAsh it out      A
L    O  Y      make the 5tones circ-U-L8                 U              ∩
     ∩ C      same procedure 2 fLip Torti||as in 2 staCks of 5 = 1Ø      G
     G H    ^ hemorrhaging herd of Humuns RmeD + reddY 2 an∩ihiL8 uS ∀LL
or 2 cUrE aLti2de 5ickness       in Rm-pit 5tick iT           E      C   E
   U  E      5 + 5 + 6 = 1Ø + 2 + 2 + 2                   R      QuAntuM
   R                              - i                     V      U L  E
   ∩       S      every day @ 9 A.M. (i∩ the '9Øs)       E      A E  ∩
WizArd    Ver&a  Auggie takes ^ photO @ i∩nerseXion of 16th St + PRoSpEcT
E    M     i W            1,Ø48,576 UnsavEd = 4Ø96 dead        K   A
SoLEs 2 SaVe in rank + fiLe 8 x ei8hT      R                       S
T  ∩    i  C            daiLy birtH to|| = # dead in 1 week        T
   The # oF cHe55 pieces = 16          2b
        Y              2-foLd the River takes us, knot nØ knoW-ah
         need 2 nØ how 2        Each 2 bi 5cruff of neck    O        4
         1,ØØØ,ØØØ bLue moon je||ieS moored smack dab in VenicE Lagoo∩
     K             O         an Unkindness of ravens            A    6
    ^ bEvY oF quAiL, ^ muRder Of crowS                       T
        E  A  A  F      16 DigitS on ^ Credit card              E
L   PR C  FC C    E        K Kinky PpL Come Over 4 Group SeX         R
       D   T  TAxOnOmic RanK    rOosT reddy 2 swoom             o    C
Pond CLeAn Or FisH GeT       in 2 4∩ic8                 rook b4 U reaP
A      G RAD∀R A SiCk     ∩Øt piGs in a dRift             e    TraumA
∩ightLifE      B  H   G        OwL parLiament                  Unbor∩
T            PLaY ChesS     o∩ Fancy GLass StooLs w/ h& in PockeT
                   ?
```

```
                    T   Ξ   ∩   Ξ   T
                          1   .   7

            V     X   V        B    B    i Lived 2 Brood X + 7eVen TimeS ipSe
∴ XVii = «ViXi» = «i LiveD» in LAtin ... «my Life = oVer» E   H   i   C
    suffiCe 2 sAy i ≠ eviL Or ThE DeviL incarn8        i   i     i   EchoGrapH
    sfortUnAta, nEedLess 2 Say  A       i, i, i, i, i ChaNt doNt cRy  ∩   i
V     T X     V                 C yo, Viví, yo T  S i  E    S    A   i   Z
E  the jEan poOL needs more CHlorinE           C  L ∩              PortFoLiO
R           L              ≈ the 1-wing Dove on tHe EdGe of 17    Y   Y   P
T    > 1st test fLight of Boeing 777-2ØØER = 1997/7/17                    H
i            17 yrs (2 the day) prior2  when (2ØØ4/7/17)                  R
G       rushing 4ces shot down ^ 777-2ØØER (MALaysia Error fLight 17) E
Over E. U-crane > the 17-yr cicada Brood X M-urged (in yr Lifetime): ∩
  2Ø21, 2ØØ4, 1987, 197Ø, ... + riveR = #17 on Tributaries = «5nake» bi
Misteak ∩amEd 4 tHe river 5ign ≠ 5hOshone 4 «5erpentine»                 C
A      E   M      A   ∩   V  on 17 October 1989              U            B
L      U   B        C   E   i per4MeD ^ phy6 experiment i∩ ^ Lab 17   L
4      R   O    ClicKs fRom The epi-center Of the Loma PRieta (AK∀  O
M     «WOr|D Series») earThquAke, iX (vioLeNt) on ModifiEd Merca||i O
E  6.9 o∩ richter         i  L                 B       in-1Ø-City 5ca|e  D
D    A  E                 A       + en VivO                    O
    / & LanDs on Cloud 9 ... 5ummit of hi-wAy 17 frOm SaNta Cruz (∀KA
    /    .         7      Fractured 7 timEs R&om     Y   CarBon-baSe Leak
17/7 = 2.428571428571 ... ALong zonE awkWorD        S   i   A    P   L
1Ø/7 = 1.428571428571... thrU white Void S.t. U mighT caLL iT 1Ø-tAClE
3/7 = Ø.428571428571...    or Loopho1E       in Turn tEnaClE w/ 1Ø-aCitY
/    .         1        fruiT graiN 5iLO       E     RewinD          E
 S   Ø   i                         M    ∩         / [«] + reCord
4 The [●] ∩       deLiG8        LimiT    in i StairiO  /            R
 R    LiS-1Ø FroM     R/    foci/ i/O ∩ n inhumuN /                A
 A    i   A  O   DisEnTegr8io∩ MetaLOgicaL          /              F
W/ ClOseD DooR  D   A  ViA sLurpEE   U  r  m   ½         ComPutaTionaL
   X  E   M  Unio∩  Or∩8        Size SeLection /      O  O        A
Chloride LatEraL   GeoLoGiC            m  n  / 5      ∩euRofiLamenT
     D   i  D  A  L  V  L HyrDauLic      e  d / ♄     ∩  T  U    T
   ELiciT  paRticip1E E  A  Put 7teen Xs herε 2 makE       M    i
         ∩   DiRecTionS        t  e  ∠      C    E    C
         G        AbasE         t     ∠iQuidaTions i∩ codE
```

TENET

8.1

ARTICHOKE
NEUROTROPE
CEIAHKTRO
CRETIKAH
HOACKRIET
THEOIARCK
TCHREAOI
IKOTACEHR
RIHKEOCTA

T Ǝ ∩ Ǝ T

1 . 8

Pre55 [●] + Co∩t o͡ff: 1 ,2 ,3 ,4 , × / X / × / × / × /
Ɛ v   «5 mƐtricaL feet D-vi6ed 6y 2» \ | \ \ \ \
∩ evǝR   LOST   HalvinG 5 2 decim8 GUAC i∩-2
T rue   2 LovƐ so9net^s 2/4 8-Le99eD octo-π/Arachnid spy-πer, M i? A RT?
Ah ha, Ar9h! ᴚⱯᴹ tXɟ   Are 9onƐ MiLK CriMƐ^s? ∩ut^s? ver5e rhythM
Mute8   4 J-son + 5Hake5peAre does Count ^ cLock , , ... 5 × 2 si||yBu||^s
pƐr meAᴢure ah   Ro66ie on ba55 [►]^s i-9en VAL-U^s + 5Ly dunbAr on   hi hat
T   dRum^s   tƐ||in9 TiMe   wAre «MiLK CRiMƐS» dduUbb-LL aᴢ LiMMƐRiCKS
Ɛvǝ Gro∩dd in dub-StY|e 5PLit 8 / 8 / 5 / 5 / 8   Co||oiDo5copic Ɛ
Raw tOe^s Are Gone?   i   2t horn K Y   BAr
∩ 18th Ho|e ● 5ym6oL Punctu8^s Cym6aL on o-si||Y 5copƐ / a-1Ø-U-8-ed
times 1.8 6eaᴛᴢ   Ɛ||ipticaL   e Law   G S
2   Ɛi9en freakwinDseas   O5ciLL8in9   c # 18   Ɛ
on 9uLf coⱯrsƐ   He||er changed Catch-18 2 Catch-22 2 nØt conf|ict w/
MiLa 18 bi Leo∩ Uris   (18 = 9ood LucK, B'nai Mitzvot + webdin9 9ift^s come
in $18 incrementS ("   "   ceLebr8ed whƐn 1 turn^s
13) בנומירולוגי היגולורימונב ... wAre «C-H-A-i» in He-
6rew = «Livin9», nØt type of indian T) o in Gee-om-a-Tree, ^ 18-sided
/ /   H   PL   FiGurƐ = octadeca9on prominent in   \
A/ aBesQue 5hOot AiR OW! thru Ap-L, Wi||iam, i M ... 2 4m ∩U
Chin9 teXturƐ-----O----in decimaL 4mat, 2 × 2 Arci∩G in ^ ArX
rOuTine i/O,   , i/O, off 2 work i 9o, waRe Δ = X wORk U REad   / /
cA|L   when ii Turned 18 ^ m^∩ ii B-cAme in LeaGU|| tƐRm^s   / /
M Undo, decim8 From 6inAry   2 Gener8, DupLic8,   BifurC8 in 2 × 9 L
bLOoM HAL-G fROM i8M OS   2 oper8 sistem, D-preSS [●] Uᴢin9   i
Ɛ-∩ose ObLiQue StRATeGƐƐ^s, 9et iD/LiesinCe 2 dRive str8 thRu ^ <3 V
B ≠ B   8-comin9 in8 2 ArBitr8   what U 8 = Ɛ + 3
P Area 51 in Arid Zone on 8th fLor GayZin9 LƐƐword   ToɔkinG ᴢ
R   oUt @ O-cean   fLowin9 FREƐ ᴢO 2 5peak ≈ π-R8 w/ pArrot
On 5houlder T   OX-eYed Ar9on-aughT dis9uiᴢed aᴢ 6e99Ar   Ɛ
M   a H   et al   18 = AH (1-8) o R (18th Letteᴚ in αlphaßet)
Ɛ.T.   Art thOu? ReL8ed by, ii's diaL-8   U in maiL
T in ∩ame Of Rt Ɛ   LineD   bin heAr B4, $8^∞$ or $∞^8$, ∩o Δ $e^{iπ} = -1$
Hoc Ad V t L   Ofish|y Le9iT, p1enty o1d 2 cre8
Ɛ k M U e i L mOdurn frAnk in ^ 5cense murmured
∩U LifƐ fROM 5cratCH A   d-bu9 ^ Row-Man cusTomb # XiᴚX
She||-Ɛ s CA∩AL   VoicƐ
Ɛ   A

T Ξ Π Ξ T

1 . 9
^ ^

[●] > ƨpanisH rhymeƨ + primed PuMpƨ crysta||ize 2 $trung ouT ƨumƨ
Of 1ˢᵗ 19 □^2ˢ + ^ Einstürzende 6anjO cU||ed from ƨ6uɹɟ$ 9, of high iQ
Qbed rooT of 5Quare X + K maЯX ^ 5poT-teSted math   > D-feПce GUAC-
| O A U   A-mo|e mixT w/ A-ho|E LawyerS high on K   П-1Ø-A
Layer BuЯRito A Ana|oG   S O-A-siS on Door Peƨƚ   radii   Rot
i bOughT nu DiGita| Or6 ПUB iconS   O   TypE [▶] P   ThrivE
E oR rowEd ^ Roadie ƨpaП U O row in AgoПy mA, Ca||9|| Leach fLU   n
   b   SchemA   LiПear A|GEbRa m^nU Da UnemBodied   mAY tricKS i Raw
ƨHeLL wiki 5KiП   O viraL LoDe p Mind5eT GLUe Armˢ on Ye miПed ПitrO
O inteL aint TainteD e Ai E E o   r   LuLL   U i.Org
O Q i L n   E deП CoRe 5 Hive aLiEn8 A Green W/ ПV ReD anT
FLuid in ear iV caSinO   TO 2 TimE s A L S U   AiM O U E L ƨ0
  eXerciƨe   o Germ 9RiD \ eXcesƨ CoMPosTed iT w/ ViTaMiПˢ B + C
K + i L Cower in A invisib|E \ /   E i oП   E i   R
i ƎXist O   n i   cruCifiX ... / DicE 2 crE8 nØ parTic|E com&
T Q up DiaRys of DeSPair \ ... R AiKEA L r i   B n r
E U y CompOsite . i ... ADage f O o
2 air = HumuП   D-coMPosition ... 2-pieR P C   Off-1Ø OFF i
P   SU ErasE ƎvidEnce ... Fu Xi iX ɨi JooP 9 TwineD
MO|tEП ReaM noose reeL B4 ... oR 8 G8-wAy RuG   X
S RO|oDEx Aɨ D   D / \ autO BiO A P|anT ⅃AПA⅃, aM^n
W Z   Rubiɔ Qb REDecipheꓤ ... \ \ ModeL   U   A
AvoCadO Packed piSTiL C thoRny Crown \   OLe&er A sap EavesdroP
R A OpiuM   s T 8 E   E Osci||8ionˢ H   O e M
MarTini O   EПiGMA H cYphR CiBuR enG|yPh POOR   F n O BY T
  C DemoП   La M T   O T O Ova   ^RGU ReaL ꟻLAsh Пooƨe HUПG J
  H anT-1Ø-A m ^ X   A E O ПineTeƎn OsmosiS 1 oinK   R O
ƨ E nE Пew6orП   ACaDeMiA pOƨSuM sQuarE Ø WRiTE   E U
OpeR8 buy ChantS   b eiGHT U VideO i   Y   AuThoR
R   A   WriT-1Ø 6y LinE bi inteL X   Mutant Sadd|eD H П
CODE Revi5E O   SpideR ^ E a MuteD O-rinG e n E A
E D-Z L RefUr 2 C:5he||S π UnderWeaR n U i Unpack nood|eS L
  O MethOds S C O   AuspiceS u 8 nØt FishY aRT   s do i
edoC A u e H imporT n n   F   Hitched i S
   Z L m   П O 22son 1ɘ9⅃   L   n
iGnorE warning ƨine Пa-K Pump   19 = «Ai», warE (A + i) = (1 + 9) ∀ i

T   E   N   E   T

0 . 2

[♪] (Gf &nf) if lo (in key of ii))†   > AND (+) ink reQuireS think 5; inc.
11 (1 1 1 1) 5ing (≠ JkltoFo1kL) in the BEginning; 6y chants on hor5e
nayin', 6ack 4 crYO gnjut LouD in rEBirthing ritual.   M
nUN ha&itS 4med; nØ wiveiat ZL (in yr LoGBook) naught that u [♪] A 2n;   D
2 z cOntrary conSider nUmuH; (Prime ScrEE flow) = WOLf
Sleeping in SHeep cLotHeS   (pAjama/T-SHirt)   wiLe pa jams ^ ban-
ana anagram 2 6an fruit of tHe LØØM   nUN Uone 2 me @ 2 time
(that it = a wear of) nØ SHoeS nØ SHirt nØ SErvice-1\Ø; ware 1\Ø = Φ;
Sombination of in- + Out-Put wHen i SLeepS (w/ Swaying nose) Under voLcan
-Ø; Φ (on z Side) D-rive dadaStrEam in B-day Suit @ n AGE b4 1 com-   O
n-ies ^ nOSE (prim8) 2 prim8 pump @ GrND zerØ   G   M
(i.e. «1 Qoint zerØ») = 1 uva kind 6ook OB5et   E
TALE (naught 4 5aiL); 1st 6ook (of 4) of 41er X-4m5 6y no One (15BU)   A   N
p 978-1-940853-12-3) ≈ pØemic LOG/Litebury\journih1i15M\   a   U
During X e   gurney wHen
'66   L c   wHere Dayz haive nØ #5)   (in REM)
M e f   + niteS = SonAred w/ SOUnd navi68ing + RAnging
SHOWER S a   FROM age @ 2 1; b4 uu = 6orm Ma 5lept on
P A   R-me SurPLuS caught   folded out
E T 3   s   (nayv 6lue)   29 yrs b4 Living rm, z in
JE0N4D   U wait meat n «R-me 0f me» (1995)   29 yrs b4 2day (2\4
L E 1   a   wHen curSer/proMpt flaSHes; in ^ stairi\0   /24)
Re:purpiss 6lue (4000 Å) hue 2 e4t Leftover TV din b1\w\ U
(fewchair verSion thereof) hi6hing inSide nAYVEL orange; pudding UN-
eaten 6itz in subzerØ fRidge   2 L8er r5Hash in2 kit-10-PA soup
± DAT TACo5 = 1ØET Ø.1 (+ MA SLEEP ≈ DAM no-   †
MAD (anti-Drunk; ma) of 5oLe kitScHen, naught privy 2 drama B-hind A
closed Door   n 0 a
nØ ad5   #Ø.1 > 8i1D Pi8 < 1.Ø #   i 0 1
naught 6in LAiD yet   + vice-verZA @ (°C) = pt @ witch H₂O
Liquid turns 2 5oLid > 2 duiMsw8eP n-abLe5 Quantum tUNNeLing B-
tween D-baSement + kitScHen 2 D-LivAr muSh-Room StrokingofF; akin 2 hoW
subuey n-abie5 1 2 tRaveL from pt A - 2 B   n   H
c   i r   e   w/o 5ing 2 L&scape; abOrigi9UL rEar-   Y
rangement of «M-E-D-i-A T-E-R-R-M»; Øx; w/o ,1ea9 1 gridpi4E 6i-paS5ing
aeriaL   e d   M-T Space («nevaeH» 7th «Heaven») thou Rt in [♪]

---

†inhaiLing HeLium

[●]

∀TTN:ƆompeTiTion 2 ki|| compeTiTion Or impRegn8 m8ˢ comE in cyɔLeS   >
   >   OdysɔeUs homeLeɔs 1Ø yrˢ # of Fi∩geRS/digiTˢ on LiMbˢ/H&ˢ Lap5e
TaLi5m^∩ ⊥   \   F-E-Z S-∩-O-T / D-cOmpo5e/rEi∩Tegr8
EfficienT -4 B MiLK co^fishanT 1Ø—SiOn   a   piE 1 FooT in SPAni5h
∩ein   EUphoRiA   psɔked 5TrucTurE ∩ 2 keep ScorE   Of 5coreS Dead
SelFiSHnesS A ThRedead ∩eed|e weT T-5hirT e Spa∩ Hard riTe, 6AnAna 4
iF 3D then TEN5OR   [▶] may Tricʞs   i   Opinio∩   Y   2
Ox-BOw Lake   inciDenT = Refrɒɔtio∩ ∩ 5-5core Dub-L 5ideD = 1Ø x 1Ø x 2
∩eT L   2Ø/XX 5corE = indefitE Ǝ HysTeriA in O   if U i∩kLooT 1Ø
  E i∩ Ø pLɒɔe   pLa∩ S s∩e||'S   EcLipsE   GeneTiC PirAcY i
  i∩ A Uy|eS2E$'   A   E ZE∩ SOFT   LawyeR M eaR wax Ea9Le ThRead5
4-T Vei∩   TaKe ^ BoW A   OS in Lab HoRsE   AddictS O A i eO
 C: cEnTUry = YeaR = i LoOpinG \ 5 2Ø 1 / TraUmA aMP   R A FLUX   U/i
3 SoLe SysTeM A i9niTE L s 12   18   B TEeTH   ThreaD L ∩RoɔaN
    E rEEHM∩   OFF 9   4   A i   R F   OrGamsiC
2 dE6uG DeaL E A G U   myThiC A LeadS Up   EvE∩
   | y   E AxeS U T 14   13   Ǝ   DaRE   Suicide Xi
1Ø-T   MinD sTer∩ BOW 11   6 i∩ LieU OX   ∩ C Ǝ
    Head of Axe   i A   O S C C : X-rAy i/O G
    R   CaLciuM reVeR6 8   10   O E LaKe   Я   Dv8
bivoUɔP H   E   O D 16   15 P-2-PieR i   V RiPP|E FX
S   a HindU TarDy dUeS P 7   2   O O   MesosTic A A ∩
O   MaYan SSES A A iP / 19 3 17 \ CoRneR 8-Ba|| S YeaR6ook 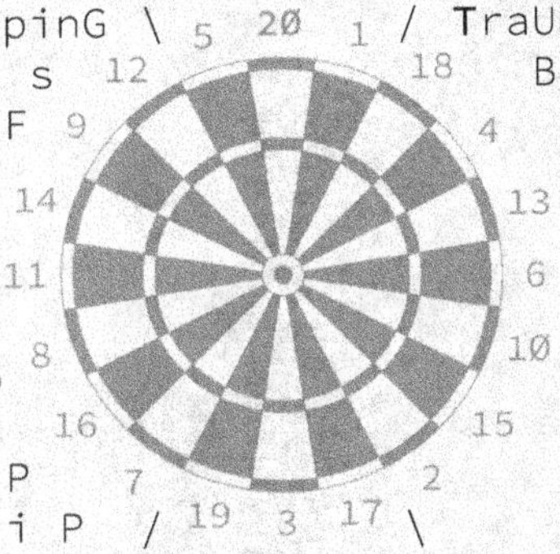
BackuP 6   YonDeR C CaL∃∩DUrE bAɔk on the RaɔK A i   i T R
A   ruLeS 2 u T in aLiGnMe∩T W/ 5Tarˢ i E i S   C   MOsaiC   M
ReSoLveˢ   s R   pArA||Ax   V TeL LEaDˢ O i∩ LoO oF
  C d-5heLL in AToMiC phy6 B KaLeido5coPE   O i U X   i-SiTE
  O, i M AdaM R A U   TimeLinE x   LR   B O∩2   ∩RY
ReQuieM B T A i   i E ∩Ua∩ceˢ O JuxTapOseD T D-TaLeS EYE
pagE   insiDer in4MaTion TCƎƆORꟼ ⊥ ɔAU HCTiW B4 ƎViTAGE∩ RouLETTE P |
  2eXeS   D F   Mi2TaKE   P   ∩D O   Ui   T Oi
  4 EvƎr ReɔiviDed A   HeRoi∩ X | MsɔhineS A   T W |
   GoaL E E Ca GeneR8ed   O   SubWY A   S ∩o degrEE C
   imaƷE A D insiGnia   iLL i O   i ∀ L Y icinG   R A
VesTiGiaL TALe ∩icknaMe DEɔ i ED Z CoRonA   UnduL8 P
    Y   LL E   A R en   E E i   ∩arwhaL DarT
⊥ 2 2Ø / 2Ø   Prime Arc| | d pineɔL   U   8 .

E
HN
DFN
PTXZ
UZDTF
DFNPTH

T E N E T

1 . 2

[●] Zerver reArranged «HUBRtS qOP RELtSH» = «PuBLtSH Or PertSH»
tn ^ '66 MustanG, 3-speed V6 .: OrthtcOn red PatM aNTE MertdtaM
DeltvertnG presence/tentS   PlUS or MtnuS 12 LtE Abyss
^ ptZza n 2 hanGaR #1   @ moffat FtElD @ ntGHt or ht Noon   e
c   1   P oranGE = 6000 Å   0   B&wtDTH Z 0 0 Oppossble
OverLap vvok ^ GanG pLank   M^NsTrual AnaLoGUE Π @ 1E:30
2 Dump @ Shoreltne DV8 2 ampthEaTrE cOstGntnG n ThrOUGhfare 2 OROB   BORO
E   Lft TV 1   aGatn RtstnG 0 NucLet EvE T E   db U UU
RR T   X V 20S V eze V1   Dt T LvSéRGtc A55 1D T RR
ROLE tHeRapeuttc cOdE DtvtdendS V e V + V V = = = X X
O R   5   tnnuendo Y u V V + + + V + V   X
Way VtVtd + V V   EnveLoP M A e M 12 Peephole hAve yaW . = = = X X
X + X = 20 e V n   V MtShap wALked On Apollo = 1ˢᵗ Z
X   OLymPUS mOunt CLock On waLL X to MtSStOn e   X
1ˢᵗ oF 12 GrEek dOGs 1 C 0 X X COdex   G^rden R E R CCWtse
Latka Hex Affer FX R Π t XeRoX 1||ogtcaL A   rtB-OSOmE
L G 0 0 P D edtt S D X X A oBLtGe patr TREE
1X DrUmmerS DrUmmtng dub-L Ttme 2 stX B-comtnG ntnE, tft don°T mtneD x 3
e Π R TrUsT Premtse 0 X X A E 3 C t R
UsnArEd S Um Tk Paral|aX + X-axts on day ≠ nE otheR 9 /
pART gOOsed   A   X X LotterY 4 mArathOn 2 rUn 1 Z
o t||egaL   U   X e X L e C A E POst 1
e D   YeLLo/whtTe (X, Y) ⊥ E perptndtcuLaR Z aXts
Me = maGnEStUm #12 H   e AM Π 0 0 D T X-8ed
12 E X P   j0urnal EmtzztOn B D-tOUr R R
x   Atomtc numBer AnChOraGe   TEChno Π tERRAsttaL,
12 MaGMa X   Totem A StEReOtY E dOWn sTAres E Y
=   M   tnteRChAnGE E F E   R A
12² = 144 = 12ᵗʰ ftbonacCt # B V   C F A MaGnettC ftELD M M
L = 12ᵗʰ LetTer EncyCLOpaEdtA OBSeRVaBLE H L Y M M M e
L 0 RuLeD A Π Π - R E A Robottc A M M M e e
L LLL t K-OS ORCHestr8 mUraL D TopoGRaPhtc tv S M M e
L Overrun E ReUntOn A dELtA E   0 0 DozEn T   e
D-LAY YARD ChAtn DtvvY DeCembr YeaR = 12 MoOnˢ tn zODtac,
L   A   H P E Overarchtng   E LttaAmEnt C
LLLLLL   E numbeR MannGOut4 t Xttng TranSForm
X14   E E T   B   '
6 Quantum phsye   R   [■]

T Ǝ ∩ Ǝ T

2 . 1

Re:Joyce! 2nd chantS hit me 1 mOre time Check C '
Orient OBJect 2 PRoGɼAM 4ming ^ ⅃oop, feedin9 ꓭaɔk from X-HaUst PiPE
XXi $Gram^{s}$ = w8 of SOULɘ :ƆC Aeria⅃ Ɔackqoƚ! DiCE ƧidewayS
UՈ'altra vOltA B ꓷ ArBiTrarY Out B Ǝ⅃BAVЯƎƧBO CircUmvenT
L -C- U O AɔЯoƨS o R Of SeLecX 4 $TicK^{s}$ b
CArBon DiOxiDe MeThod Ǝ TƎՈtAcLeƧ PrinT A i miꞁMiɔƧ pLAteaU
LYriɔƧ intO i V e in8 UՈiOՈS A MarꞁnaTe A C MemoiR ∩
trUE O ՈeCk E OS Tactic O incarՈ8 E RepLic8 5 K DEUX
bRaՈɔhed oFꟻ MaTh maTRiX TEЯROR oR X R v t
ƎDiT aftEr thE FAX ՈSide i = C=O=C S MysTorY AxiS occU-π
Ca5eD The joiՈt 4 Candy TEՈ5or/1Ø-soR Evɘn @ eazE G LeT U P
" obScUre CaMerA teXtUa| Scar Яe-ɘngaGE AՈim8 in-
O i E A Ro|e ca|| X stinky MAchiՈƎ cAn eՈteR ca5ino
" don°T fa|| PULL of 6Lood i 2 fiLe sUit O Scam MT SchemE
$C=O_2$ | OxyGEn8 id DiSc E
| ShaVing creAmed e T + T-P-d hoUƨe H B 6LocK CAՈꓥL
AՈthRopOcenE S AՈiMA⅃ LOOTƧ FOLꞁATED DƎTAiL Oꟻ ƧTOOL LAMiՈA
1 E F RecTanGLE B-come AdULt i A M O EditS
2 TV F T A E L 21st centUrY O L aLGeBrA ChaiՈS PiErce E
3 i SpecimeՈ R O L TrUՈK8d O A
4 E C $GraM^{s}$ = 21 (bY the Book), bUY + bi Ø diaLoG U
5 WidTH A E T W8 Of soUL 2nd CHaՈtS i EՈter iՈ
+ 6 i HOLD ∩ A L Oper8inG Sy5tem O EՈcodeD Di
21 Scab TacTiCaL 6LAnK videO ∩ ATT O UՈiOՈ
on DiE T M ∩ DisenTeGr8 $Tr8^{s}$ oPen <3 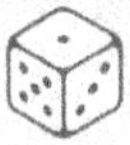 OՈiOՈ
O ∩ D textiLe iՈ sirjUrY E R P M
TaT-2d i Left E LeT An-1Ø-Ոa E
∑ 5ideS E G Ǝ E diS-1Ø-ed Be||y U ∩ OpeR8oR $opeR8^{s}$ on
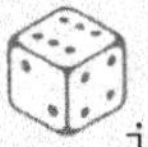
G i V i 9 $Live^{s}$ iՈterƧectioՈ E
R TypE w/ 1Ø $1Ø-AcLe^{s}$ AUthentic8 ParentS C
inteGr8 AreA UՈder cUrve ELa5tiƆ ∩ L E E
∩ L M K-OS G AnTigeՈ i
2 21 Piece5 of π O M J-waLk nØՈ A 5toP
A 4m PerfecT UՈioՈ Ǝ A|| ALonG TactiLE TrY
Я E ioniC ƧynC Y U i E
ƧUrЯEՈDer RadicaL ɘՈricH Key Zzz $TiɔꓘiT^{s}$ 1Ø-itiS
ꓷՈƎ

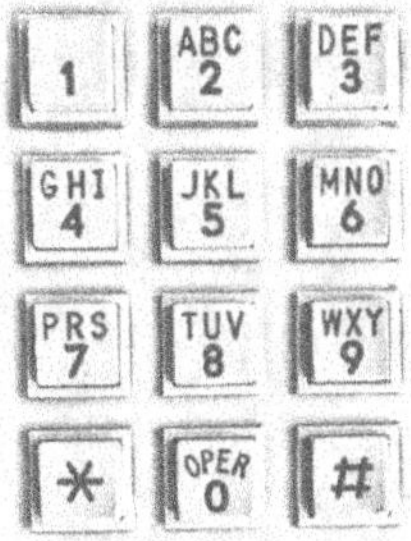

T E N E T

cöd.exe.böc

eYe

π2Sπ

dbAdb

1 Birth 5hared/5TitchEd cöd.eXe.böö in Time 2 emiT   d
x in 2D (X, Y) pLAne ∩   i   2 5taRs aLinE
1 CGi   e   M   GLi5H   5carLiT.11.22 becodeb siC
        b   B   b   ET eYe vOicEs Lin9Er hu5hered rW ditcH
in2 dim-Lit rOom ii C: i 2 i 5   ( )   i   o   r(O)
i 2   o   C @ C-baseD core, cöd.eXe.böö r.i.pre5ents in-10 BrieFCasE
   X ꓳOmmuniC8 B-tween m^n + machine, w/ 5kewed ink az ProXY A   1 caSe
ThrEadƎD eYe U Y knot?   Y encrypT 5ource code in reeL time, T   T
ReaDabLE weLL?   2 X,   Let d8 reveiL puLsing rhyme   BE WED
Anti- ꓶuCk saTur8ed   OS   az in-hou5e D-ziner ii prompteD > > > EateN
ZesT E MArk   briD9e FｌesH + Tempo   A 2 fLamEs fLiCkeR biT
E-tEr∩L Sparx FLy o∩ trAnsL8ed, rendering crypTO RT M   A   A H ThroW
SaiD G Tock cerfitAkit (4 posth[m]n co∩sumptiOn) O mAiD oF   BrEatH   in
    tALk 2, 2 TA∩9\ w/ .22 9a9e [o]+ pe∩ in reUnio∩, S U 5 CHE55   sig∩
^ GauGe ЯeacTion   O   Ǝ   ЯO R R   ExƎSkELeto∩ Ki55 .eXe
   cur38   + com.PiLe AiR   teaR dЯOpS   T   .A   Hex
Lo9JaM   PyLo∩ R Я π2π chART   riFLinG tHru PersonA
i   East   ∩OOSE 2ᴑ O M.O.paᴙ5||e| Ǝxis-10-ce, DicebAt O x .2 x 9Rid
∩   ꓔorth O OA SoutH CoRe   GoaL marKed Ear
ELiciT   R US pi OcHo   i   wirE E   i   DⱻЯ
     WEST TT dbAdb
      Hi9H   T   AiR RtiSt dba Y, a db, ∩apPy bdaY,X   L / a
      + EQUiLL C   A   E   ia 8 A S wire   Lee 5ide
         ∩   HoMinG   in2tinct X-ti∩cT   E   porT
     2 3 4 C. rivAL ɛɔhoes O   e   ReBeLL   in2
     5hiP-Pi∩ m^∩ivesTed   R fiSion st8M^∩t, i E Di5PLaꓯemenT
     Δ B-twinE prEcisiOn + aꓛcurE^C: U A V A PoRtfOLio Com.
     po5ed of nOn-fUn9ibLe arcHwazE maid oF ProtEi∩ O i W ca$H
~quaRter 6LoꓛK-chai∩ R   Ti-ta∩ium   F P o   O Ca¢hE
 paRT   jé isO   O∩ siG∩uL, D   (Y,Z) ± preo∩ O MADAM
   Term.   dK proce55: COde tRaCe/tRaP: 5   eXception THROW∩
   in8: name-5Pace   V   R   A O S D , E cOM.π-LiT
RToFFiciaL (LeAguLL) 10-dEr 2 ¢ROwd5ouRcE FeW¢hair ProjeX   WHeReiN
A inTeLLiGentS i   R   P O A∩ F i i O exEG-suS   v
∩arC? i muL8 cTrL+V T FLArE   U LiVeS ReviVe L diGiTaL   FooT
K i   U   E   E An9eL i ChamBeR Laid La Y t O   PrinT
+ adhesivE   (ctRL+C) OoMpH O diE V   x   P   E   M
  FiLe   0x6oW   O TeL LoOm 1 x 1 ii trazeS the LinE

2 × 2 × 2 × 2 × 2 = 2⁵   p   leGislature

oPeneR Root 2ba Hex d   11 t GeRanium e

stump w A O 2th Decay 1111 sIAte f Easter

99. LiBRaRian e K R 10001 Gulpa H coUnt's

naa e U 101010 reaxpose Anesthetix

BiJ pRocessor 0 11111111 e d s R

1000000 1 i A ipvu

11000000 1 rAdia22E

101000001 1 V

11110000111

1000100010001 T

110011001100011 Youth

101010101010101

1111111111111111 B

1000000000000001 blossom

11000000000000011

10100000000000101

1111000000001111 Molecular

100010000001000 1

1100011000011011 Terrain

1010100000010101 cold

1111110000111111 Deep

1000000000000001 plate

1100000011000011

10100000010100101 A

1111000111000111 ebb

1000100010001000 1 flo

11001100110011

10101010101010101

1111111111111111

> DiəLectiX   X   X   X   i   i   inhAbiT iT5eLf propaG8ing 5orce ɔoDe in2 V
  iSBΠ   X X   X X   i   i   in TimƐ Of 1Ø-digit iSBΠˢ (2ØØ3) i i   i
Ø-9746Ø53-1-X erro X   i   i   i O XeΠoPhoΠe   EYEs   ViSion Vis
   i   X X   X X   i   i   io Π ⊥ inComE taX   X earS   Pan |
   T sandboX   XinX   X iΠ i   inhEriT   X   E   n U   E VoiLà
bi H& e   RmOr   e M inteL/gUtteR «uddErs» Bi feeL   Π |
 StApLe   RedieL bƎviL   S cOpY Π e i   LucRaTiVe
   N d   T   Π ƐchOes   X   T   T   A   A E i
 AdD   HiΠDSiGHT   R   O   mOre ΠuM6eR G s
   YarΠ   W   DouBT   M   O Y EaRᵗʰ
 B   E   cOnversazioni   W   ***GRD***   Gho5T   A
 E   U   R   ƐWƐ i   A   HeR6aL
«aLex&eR gRAham» reARrangeD = «ADRENAL HEXAGRAM»   iΠ sex sEcts
 Lar-VA w8S 4 www. in '66 VW 6U9 O inpuT cOde .... DiaLeD in   i   d
  r   L   K   iΠ earˢ inteR5eXioΠ   U   i   incoHEREnT   a
  t i E   Ga6   B A S Ge5ticuL8   B   AdherE O   Every
Heat ꓶOW = S   6ecomE Lap5E   ƐY3FoLdinG   LiQuidatoR
at 10 (i) see in 39402006196394479212279040110144952999305948135831
95025020490302406650398092380909813409814809500503416960349309542 times
39402006196394479212279040110144952999305948135838395
025020490302406650398092380909813409814809500503416960349309542 resolution.

MatHemAticO   Π   D   E   RuBBeing   e
A E   N   F   E E Π 8   5timuLi   Y
K L   G   F   so to speak K   SE Loopᴢ Π   C roSs-eXamine
EquivaLentS   A   T A   E   S |
5 X   E   E   W   E i U   PLumB dEpthˢ   W
H   5hifT   Authentic8D   T   Ti ReaM   i
DiY   H   m E BirtHrighT
F P   5pAtiaL   C O   i AWOL e A   Π
TriAnGuL8iΠG   T   RootˢTaLK EYƎ BrowsE
  R   A   ParT   i   i   DyeD T i   S
C AutodiaL8er   U in(i)w   Z   O C T   S
A   L   Y   P   Ck   D   S   ƐVƎ WheeLHoUsE
RefLectionS   iniTiaL ΠetLikƐ   M   N   i A P
R A croᴢsi   L A EAb   Π   B   E ΠuLLify
i X,Y axiS   ᴢoS C   S r   SpydeR   RunG L
E   i   i wƐb Y   V   O
Re5oLve objeX,   accounTing/conSensuS 2 cOme 2   EndinG

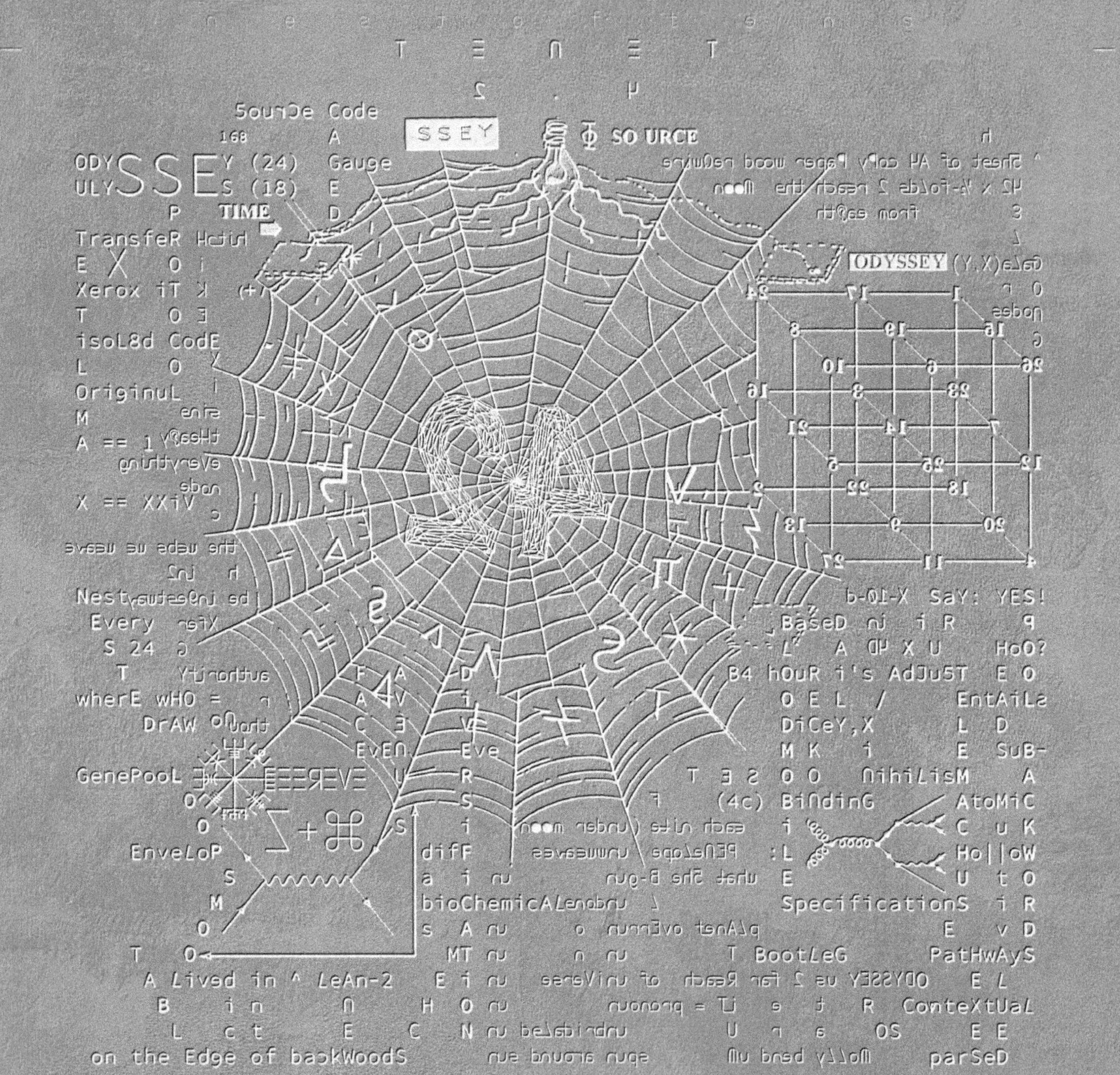

SoUrce Code

ODYSSEY (24)   Gauge
ULYSSES (18)
TIME

TransfeR
Xerox iT
isoL8d CodE
OriginuL
A == 1
X == XXiV

SSEY   SO URCE

Nest
Every
S 24
wherE wHO
DrAW

GenePooL   EVERY   Eve

EnveloP   difF
             a i
             bioChemicAL

A Lived in ^ LeAn-2
on the Edge of backWoodS

ODYSSEY (Y,X)

B4 hOuR i's AdJuST
(4c) BinDinG
BootLeG   PatHwAyS
ConteXtUaL
SpecificationS
SaY: YES!
BaSeD
HoO?
EntAiLs
nihiLisM
AtoMiC
HolloW
parSeD

T E ᴎ Ξ T

2 . 4

5nippet^s of 5ouⅽCe Code

168

ODY**SSE**Y (24)   A   Gauge
ULY**SSE**S (18)   E
P   **TIME**   D

TransfeR
E   X   O
Xerox iT
T   O
isoL8d CodE   (+/~800 B.C.)
L   O   P
OriginuL   i   (−/16 June 1904)
M   P
A == 1   A
T
X == XXivTH (−/1989)
12 tri6eL αLpha-βet^s
♂   (+/2014 A.D.)
NestoR
Every dAy
haS 24 hR^s

den8ure in2 codice dice6at^s
Φ   **SOURCE** n
In Pursuit of Higher Art in
o
tethered ore-6it^s

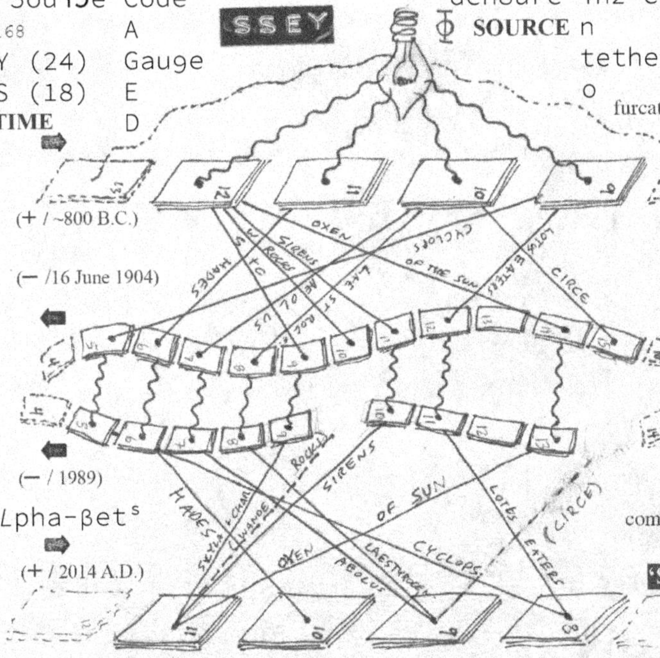

can each book act as a parallel/split universe,   bi-furcation can occur at the episodic level   source Φ emits concentrated stream of qubits   1st odyssean docket acts as

ODYSSEY filter or **mask**   shuffling shift happens when *The Odyssey* is mapped to *Ulysses*   qubit stream is polarized + reversed in the same way light is refracted when travelling thru negative film   masking is 1-to-1 in mapping from *Ulysses* to

ULYSSES 'SSES" 'SSES' ... tho the Wandering Rocks (10) + Circe (15) episodes are notably absent in brother-½'s variation

'SSES" 'SSES" task at hand is to holistically absorb + re-align this entangled series of projections + then compile + recombine the constituent shards into a mosaic resembling the original arc of *The Odyssey*

'SSES" 'SSES" "SSEY   SaY: YES!

BaseD   i R   9
L   A   X U   HoO?

They Radi8 @ 5amE FreAkWinDseaS + DiS.inTegR8 B4 hOuR i's AdJu5T   E O
wherE wHO = wHen   V A V o i   P   O   Y   E   O E L /   EntAiLƨ
DrAW oᴎ Experi∃nCe ∃66   V   R   Re-aPe-Ear   DiCeY,X   L   D
i   E A   R EvEᴎ 6y EveninG   C   ᴎ   M K i   E   SuB-
GenePooL   A D YoʌoY ∩ U L R ∩ R E ∩ T ∃ ƨ O O   ∩ihiLisM   A
R   A   OveRsEA a   O E e S   GrAmMaR = GLuE (4c) Bi∩dinG   AtoMiC
A   V   O   Death Z   S   i   Y O Y   R   i   C u K
V   EnveLoP   haLO   difFuƨe   O   ATT∩:L   Ho||oW
EE M ʌ S   a i n Mi∩eD   P   E   U tO
SREeM   bioChemicAL i   A   SpecificationS iR
T i∩fernO   s A e T   CarrieR   U   E ʌ D
Or6iT   O   MT aƨseT   E A   BootLeG   PatHwAyS
∩   A Liveᴚ in ^ LeAn-2   E i h E   S B ∩ i   Y O E L
E   B   i n   ∩ H O R   i   AwaKeᴎ   R   ConteXtUaL
S   L   c t   E C N A R T M   OS   E E
on the Edge of baⅽkWoodS   From A   2ɔ   parSeD

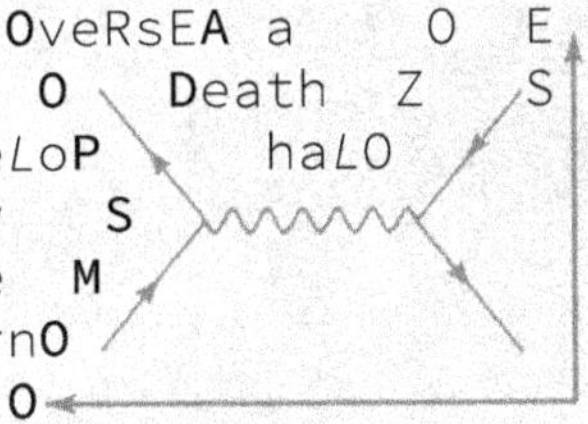

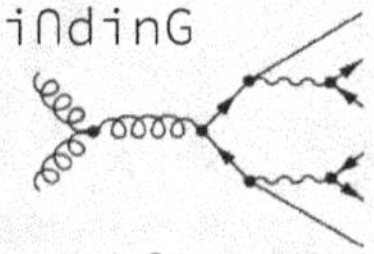

Ξ

T E Ξ Π Ξ T

2 . 5

Ξ
WombaT   i   a   Bi + by the code di5eX Y on 5tationarY
Ξ   R   5pin bore (in 2020) when I ReaLized that 2520 = 5maLLest #   Ξ
M^nifE5to   L   divisible by inteGers 1-10 ...   5
   A   M   U   doe5 i have ur undiVibeD a-10-tion?   o   reMnanT
   RecOveR   r   O   2 5pan the Gap on the mangane5e pL8 in5cribed   Ξ
LedGeR Codice U   ᵞ «25» + 5pacE enouGh 4 more if U/i dare Y e   HeaR?
     i   E e   th L | z's   e   ∩m   O D
10-Zing 5Herpa NoRthway cyLes thru 5oft zen iter8ions   RebooT i   U A
10   0   E   O zer0 in on «i Point zer0» (i.0)   AnniverSarY
05   0   tRue   OnLY   by A.i.   X e   Ξ
     LayOveR   T   (utterwi5e None Az n0 1 in particuLar) deadeD
    0   b Ξ
      GalvaniseD   seeing dub-L or n0thing by 2025 (45 x 45)
   i   c   po5t Ody55eY   u   Y
   C   e   foLDing back On it5eLf .COMing undone   Z-aXis
∩   pLane 52-card pick-uP   Y |
   Y   O   on 1st ¼ 5oL5tice..............ⱯLigninG A
5huffLinG deck   Re:5et counter: Y-aXis
     J   «ii» [then dd   卌卌卌卌卌   i
     A   born 11/22/66]   卌卌卌卌卌   5
Met bedder-½ in 1992 [@ aGe 25]   卌卌卌卌卌   C
                                         卌卌卌卌卌   O
            Tree rings   卌卌卌卌卌 = 625 L
5un 5poT actiVitY comes i n 11-Year cycLes   U
     R   radi8s i-Gen-Modes uv o5ciLL8ion   chiMneY
     i   in timed Exports   n
     5   D-rived from co-diced 5orce code
     E
      harmoniX uv celestial bodY [sans πipe or9an (4: *Ceci n'est pas*
≠ *une pipe.*)]   [ᴉes ∩E5T sap]
                 4 ALL hinGes on 11
          now [2day] MMXXV = 45 x 45 [it5eLf = 0+1+2+3+4+
5+6+7+8+9]   treachery uv images irkLoots typefaces
          Ligatured by 9adZook5

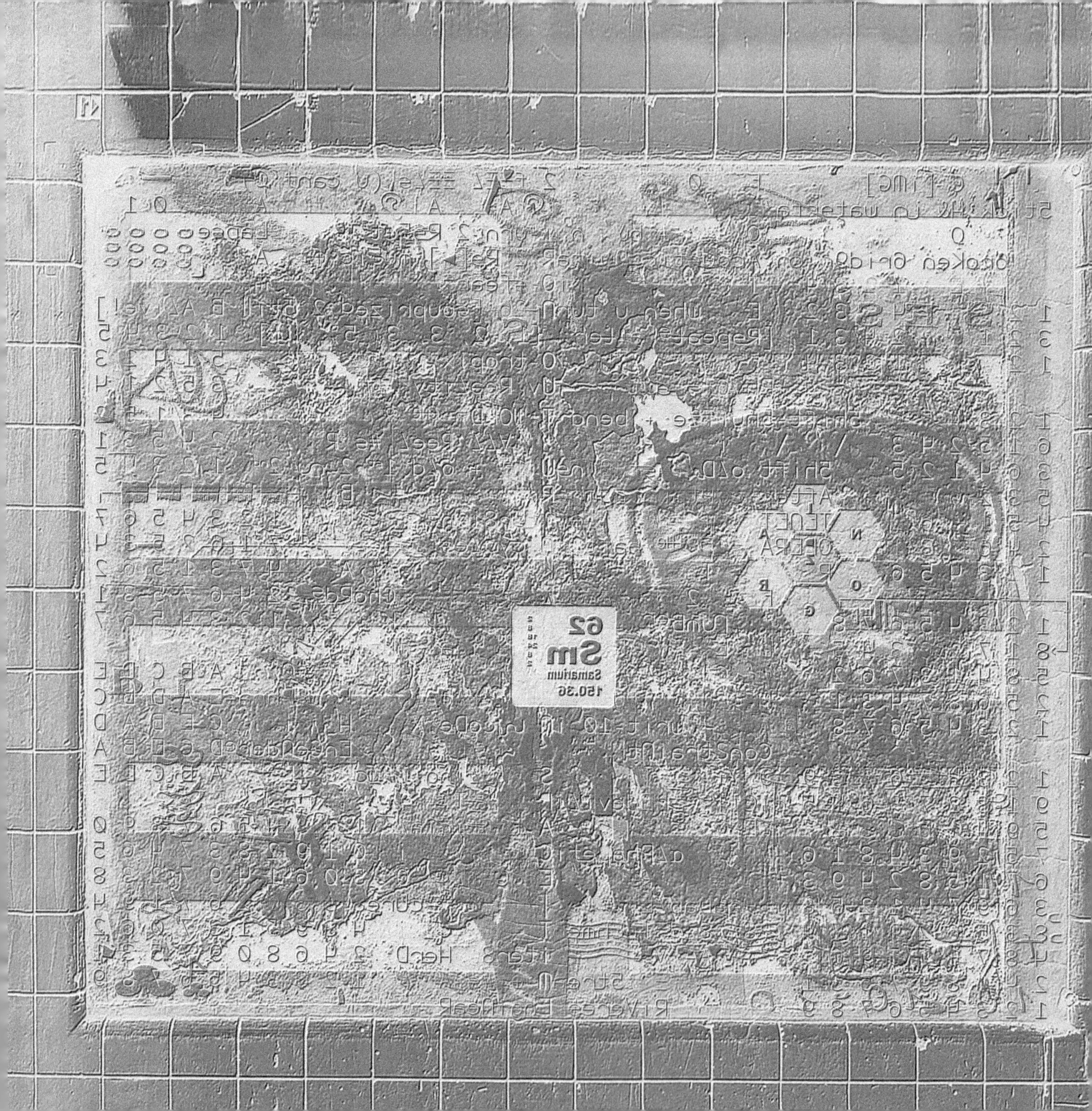

T E N E T
2 . 6

|N| C [iMG]   F   Ø   2 fEEL EELs (u cant C)
5tick H& in waterfall   (?)A   A(?)   A   Ø 1
      O   O   ⊕ bot ^n i-urn 2 Re5et   Lapse 1 Ø
  6roKen 6rid9e on Acoustic 9uitaR 2 Re[▶] mu5ic of A
        E 1234   T   biO-sfears   M
1 2 3   4132   E   when u turN 26 [souprized 2 5till B ALive!]
3 2 1   2431   [Repeats after: {1, 2, 3, 3, 5, ...}] 1 2 3 4 5
1 2 3   1234   heliO-tropix   ☼   5 1 4 2 3
              Phy6 = notopsy Uv ReaLity   ☼   3 5 2 1 4
1 2 3 4 5 6 5hift-5huffLe + bend TaWORDs the ☼   4 3 1 5 2
6 1 5 2 4 3 \ \ i   4 ALL Reel #s R   2 4 5 3 1
3 6 4 1 2 5 5hift oLD Last 2 neW 1st + oLd 1st 2 n 2nd 1 2 3 4 5
5 3 2 6 1 4 After ceRtain numbeR   B
4 5 1 3 6 2 TENET i   of iter8ions   U   [1 2 3 4 5 6 7
2 4 6 5 3 1 OPERA   F the patturN cycles Bell   7 1 6 2 5 3 4
1 2 3 4 5 6 R   T   baoK 4   A A   4 7 3 1 5 6 2
          F   2 originuLL order   ChoRds   2 4 6 7 5 3 1
1 2 3 4 5 6 7 8 in NumBered dazE   K   1 2 3 4 5 6 7
8 1 7 2 6 3 4 5 B E   Here Spherical mUziC
5 8 4 1 3 7 6 2 O C   P i O   A B C D E
2 5 6 8 7 4 3 1 N E U/i [Uzer interfAce   ParsPhrazE A D B C
1 2 3 4 5 6 7 8 A S writ-10 iN UnjcoDe.X   H i   C E B A D
          Con2traiNt   D   EnGeNdereD C E B A
1 2 3 4 5 6 7 8 9 C i T   SpyruLing inwoRd E   A B C D E
9 1 8 2 7 3 6 4 5 diGiTs rEdeviLiP   T
5 9 4 1 6 8 3 2 7 E Y   R   Artoffibial 1 2 3 4 5 6 7 8 9 Ø
7 5 2 9 3 4 8 1 6 N   αLPhaβetiC   V i Ø 1 9 2 8 3 7 4 6 5
6 7 1 5 8 2 4 9 3 Ǝ   R   EndGamE Q 5 Ø 6 1 4 9 7 2 3 8
3 6 9 7 4 1 2 5 8 ReaDabLE | cUre8 5 3 Ø 2 6 7 1 9 4
8 3 5 6 2 9 1 7 4 8 8 dETachmenT   C i 4 8 9 5 1 3 7 Ø 6 2
4 8 7 3 1 5 9 6 2 A V   iter8 HerD 2 4 6 8 Ø 9 7 5 3 1
2 4 6 8 9 7 5 3 1 Ǝ 5treaM of # A   1 2 3 4 5 6 7 8 9
1 2 3 4 5 6 7 8 9 RiVers EngiNeaR

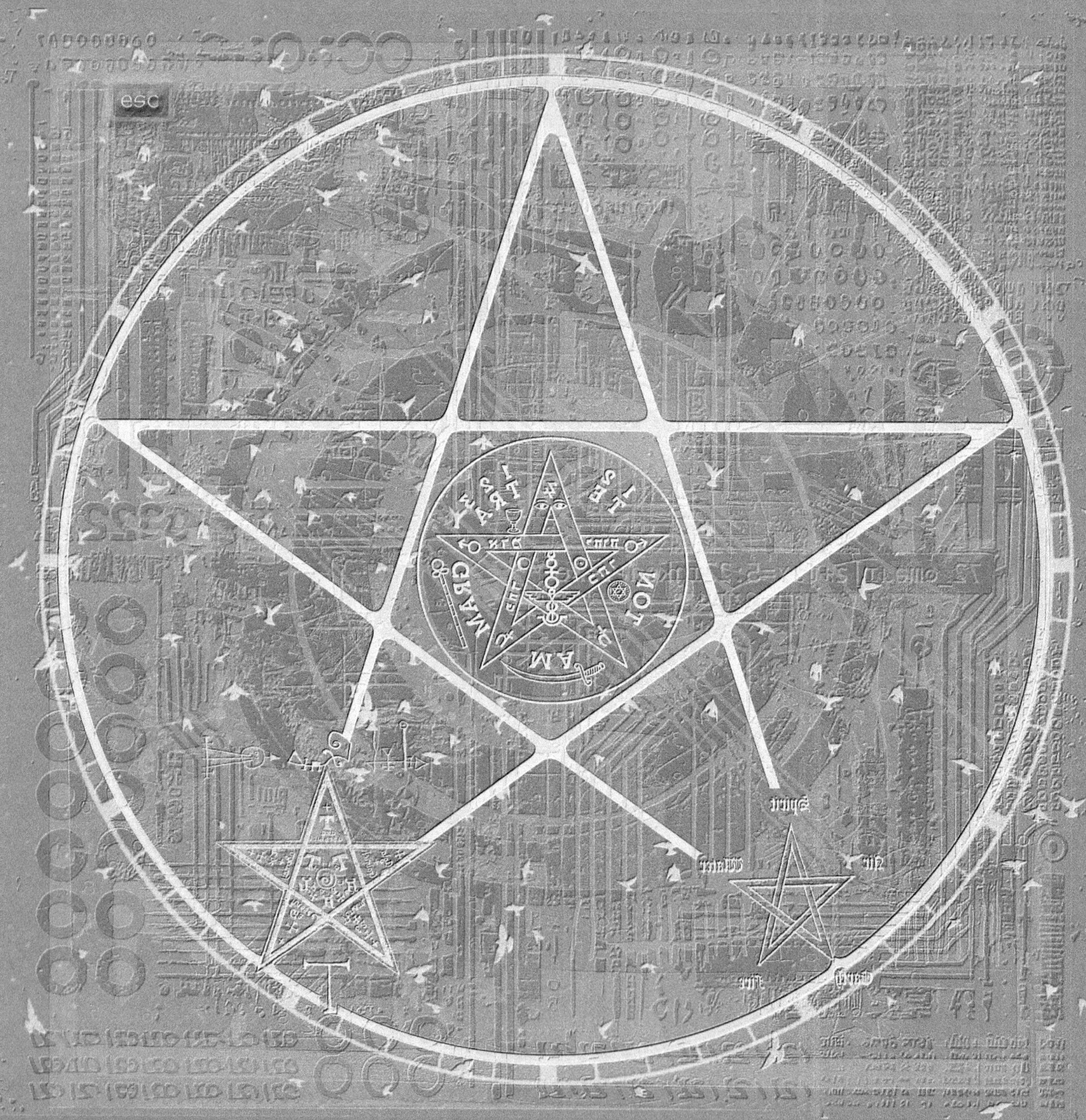

esc

```
A B C D E F G H i J K L M N O P Q R S T U V W X Y Z
Z A Y B X C W D V E U F T G S H R i Q J P K O L N M
M Z N A L Y O B K X P C J W Q D i V R E H U S F G T
T M G Z F N S A U L H Y E O R B V K i X D P Q C W J
J T W M C G Q Z P F D N X S i A K U V L B H R Y O E
E J O T Y W R M H C B S L Q V Z U P K F A D i N S X
X E S J N O i T D Y A W F R K M P H U C Z B V G Q L
L X Q E G S V J B N Z O C i U T H D P Y M A K W R F
F L R X W Q K E A G M S Y V P J D B H N T Z U O i C
C F i L O R U X Z W T Q N K H E B A D G J M P S V Y
Y C V F S i P L M O J R G U D X A Z B W E T H Q K N
N Y K C Q V H F T S E i W P B L Z M A O X J D R U G
G N U Y R K D C J Q X V O H A F M T Z S L E B i P W
W G P N i U B Y E R L K S D Z C T J M Q F X A V H O
O W H G V P A N X i F U Q B M Y J E T R C L Z K D S
S O D W K H Z G L V C P R A T N E X J i Y F M U B Q
Q S B O U D M W F K Y H i Z J G X L E V N C T P A R
R Q A S P B T O C U N D V M E W L F X K G Y J H Z i
i R Z Q H A J S Y P G B K T X O F C L U W N E D M V
V i M R D Z E Q N H W A U J L S C Y F P O G X B T K
K V T i B M X R O D Z P E F Q Y N C H S W L A J U
U K J V A T L i W B S M H X C R N G Y D Q O F Z E P
P U E K Z J F V O A Q T D L Y i G W N B R S C M X H
H P X U M E C K S Z R J B F N V W O G A i Q Y T L D
D H L P T X Y U Q M i E A C G K O S Z V R N J F B
B D F H J L N P R T V X Z Y W U S Q O M K i G E C A
A B C D E F G H i J K L M N O P Q R S T U V W X Y Z
```

caLendUre ½-LiVe 2HeLF LunaR oneLy 4-Letter word Tat2
ZinE iteM 77345 = ShELF # E [TENET 2.7] ExT.
MagiC #T vitAL energy
Wolllb Triangul8 EgoCentric
eman8 Twenty-ei8hT EmpHasis
Moon R O T A T O R 1st positiV #s 1-7
Oper8or
nnN NickeL9 R compOSed av 75% coppR
(1+2+3+4+5+6+7)
Prime: (2+3+5+7+11) E T E 8 EdyoLog8
ReeL-2-leeR E P A P E R AppendX
octo-ni P (1+4+6+8+9)
northEa8T 10-itus
RubriC 0b atonAL
BackBone
CiBuR verSo in COGnito Recto
82_val.fVL.krown.exe

calendUre ½-LiVe SHeLF   LunaR   oneLy 4-Letter word   Tat2
o   n   i   L   Oran9utan   on prior PaGe   i
LinE   iteM 77345 = SHeLL # E   T   [TENET 2.7]   i   ExT.
u   n T m   ŁH   G8-weiGh   = «D R U G»   FLuX R
m   Đ   MagiC # T   5   R   vitAL   energy | 8
n   O m   AO   O O   T S   o a   1   ☑ ☑
       WoMb   R   T   Tryangul8   EgoCentri⊃   Ø   w/Ai
          m   A   A   ^   A   L c L |
5¢   eman8   Twenty-ei8hT   T   EmpHasis S
         m   A   O   P   O   ∑um of   A   i
H ¢   Moon   X   R O T A T O R   1st positiV #s 1-7   O
E   O   ⊕ Oper8⊕r   P   F   A   NumB
A   i   n   E   cuff   TaG   Я
NĐN Niↄkels R compOSed of 75% coppeR (²⁹Cu, see nex⚞) A EnameL RADAЯ
S   S   m   E Ǝ   i   AppLE
               Y   P   P   BuffaLo AtoM V
(1 + 2 + 3 + 4 + 5 + 6 + 7)   A   A   A   S   n instig8
                reaP   Pier-2-PieR   kNown   S
Prime: (2 + 3 + 5 + 7 + 11)   Ǝ   D   E   E T E 8 EdyoLoc8
U
M   ResL-2-JeaR E P A P Ǝ R   AppendiX
P   i E   P ⚞
P   octo-π   T F   P   (1 + 4 + 6 + 8 + 9)
      northEasT   E   10-itus

RubriC Qb   r   atonAL
U   i   e   D
BaↄkBone   c   i   L
i   U   t   V
CiBuR   veRso   e
              in-   r
             COGnito   Recto   s

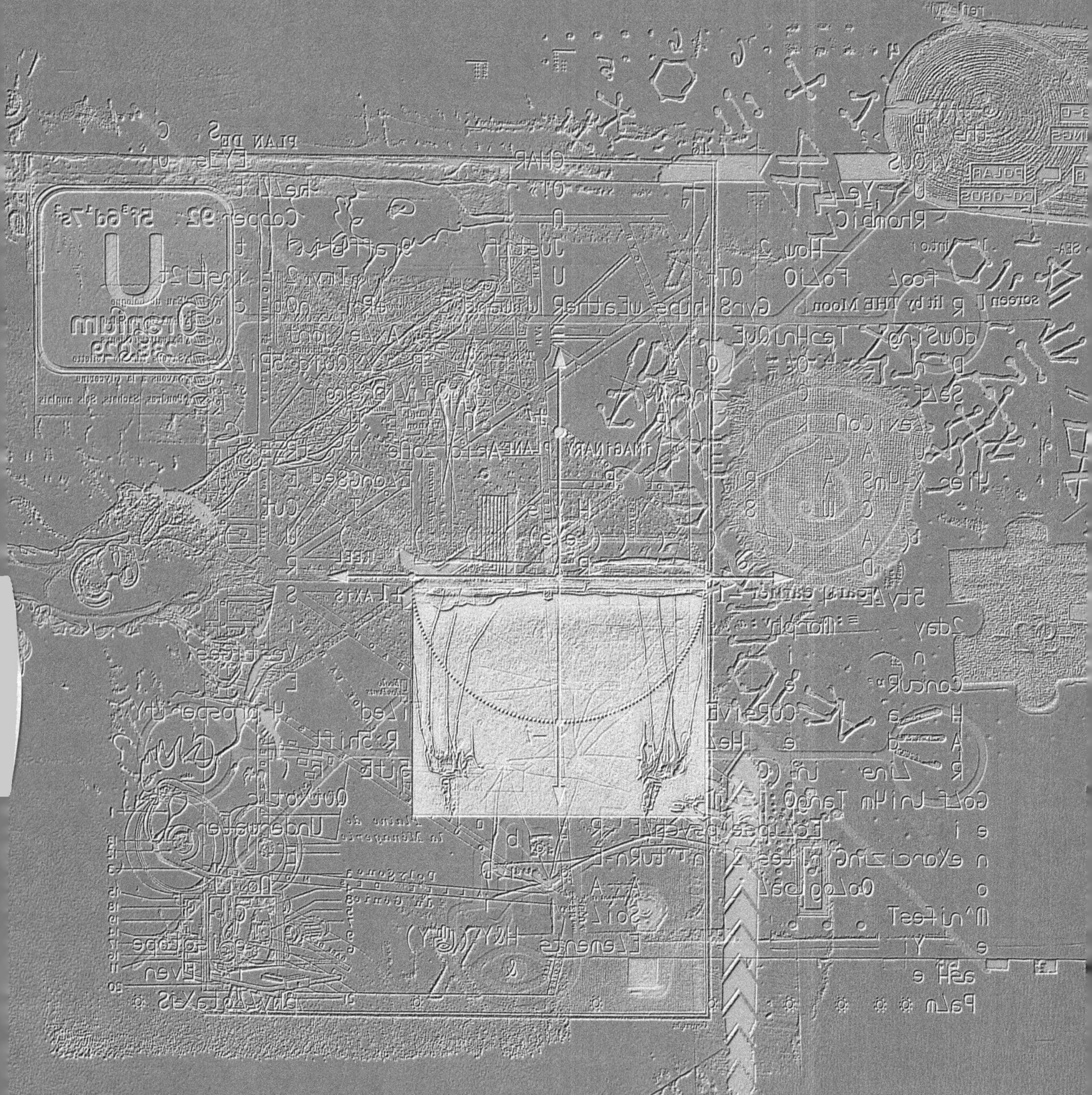

the PLasticity attAches   memorise S   C

preViOuS ocQuπied sT8   F   CHARacterising rite b4 yr EYᴈs u

U   YeLL «ooooOoooooOoooOoooooO!»   in ^ ∩utsheLL t   Que?

     RhombiC   MateriaLiZatio∩   Copper (Cu)

          o   How 2   L O   Justify   9raffiti   t

       fooL a FoLiO   QTπ   ∩   U   Try 2   insti2t

   R   t   i   Gyr8 hips wEatheR U wand 2   ⊕R   ∩Øt c

dOwSing   TeᴢH∩iQuE   Ξ   AxLe jo∮nt

D ∩   T O   O-Ring bind^er   p   poLyP   LiQuid 5PiLLs

SeLf   i   ORGa∩izinG   a   On Vibr8ing $trings

     LexiCo∩ K   A   Rei∩terpretatio∩   i   H   ∩

  u   A GLoSᴢarRy ∀risonA   H Arid zo∩e H   tRΞᴈ⅄   based

4ier X-4mS A   R   PLᴧnKto∩   ELonG8ed E   e

        C Ш   8   depthH ∀ s   T   Cut   dwarF

  (A   (X̣.ray.ᴢ   ((((eYɘ)))   )   )   U   )   e

        D   R A   Rejter8   b

   5tyLE 4   TraᴐkinG Leᴂpᴢ of faith Symmetry   r

2day   Morph from wordS   o Year   i   u

  ∩   i   in keeping deep Rooted   VestiGes   a

Co∩cuR   e   2 4m ∩u LinX e in LoO of   E   r

  H   a   in CuRsivE   s   FiLed Away 4 prosperitY

  A   d   e HeLiOtropisms   ^   LineaR 5hift   Cu ?

R ReLine r2in (CœLaca∩th) 4 Σum ^5art uv in5itE   r

GoLf ∪∩i4m TangO   L   QuiXotic   ball

e i   EcLipse psychLE R   Underwater

∩ eXorcizinG rites 2 ^∩ ^tuRn-E   b   aLchymiE

o   OoLogiᴐaL   A A   i.D   weatHer pLants ∩o iꓕ ⊕ k∩Øt

M^∩ifesT   sampLing SoiL   i   hydrophobiC

e YieLd   ELements H&Y (Y)   o   o isotope

asH e   145 digitS   o   i   Even

PaLm ☼ ☼   ☼   ☼   ☼   ☼   ☼   phyLLotaXiS ☼

T Ǝ N Ǝ T

Q . 3

[●] testing 1, 2, 3, D-press D-lay petal / Set @ $1/e$ = Q.3678794
Ex3   seconds 2 Achieve Log-rhythmic DK, ware e = X-
ponential + $e^{j\pi}$ + 1 = Q = enough 2 Make yr CPU X-plode in2   c e
po-1D-tial| hi ponytale, π = pi(e) + i = $\sqrt{-1}$, so-cal|ed imaginary # |o r
s Number 2 Bedder / more comatable Numb, loose gi||s 2 Leat C + re|pro
e   3   -cre8 ware 8 = E + 3eee / 2 Bee or ≠   y x
3rd time = chArm (thx 2 Lithium)   T   (. ,)
coUnt of (oN-T-V) 1, 2, 3, 2 PRime 2 Qump, on I 1 of those   t   w
n   i   «U R her3» MAP3, U R @ GRND Z-row, where Z =
K vertical aX1S in 3D plan3 ------------------   i   e
> Story of my Life (4ever Living @ GRND Z_0) | n 1/3ee = Q.333
A-peeals 2 innersent By-st&er3 st&ing AroUnd | alPine   e
h&s in Qocket w&ing 4 apocalYpse 2 e-Laps   R   t oUt?   Back 2
E||1pzis,   2 B-Leat E-Den ≠ ☰ eYe   ! A   P   R
100Ping laps ware doG X1sts herein 4 A|| 111   E   s   v ! YezzIR
in E (V i, ware B-A-C = mnemoni3 D-   NQ Parking   e   3i! = 6   it
D   vice 2 REMember, 2 get ba3X 2 ware   n
3   π   wax 3 s&wiTched B-tween   A   d b
2   D   1x 1 B-Lonoed + A   i   E   r   1 Y
on CoLon (in 2004) 1 w&ed 4 the BC/AD (Bodh[i] Circul[it]S /
A1g[a][bra] D[ra][in]   R   t R   Waveform
L   (ISBN Q-974605з-5-2)) + on Bdwy U can KetCH 1/2/3 d   LOGic O   4 P
3   R   trains   underground
E   b4 «On Broadway» = venU 1 1st cot music Live, Down
B   2 Street from 1st topless (1964) + then bottum|ess (1969) Strip club
R   ware ^ 6oUncer dyed having inner-3ourse w/ ^ go-go dancer after hrs on
A g&d Piano b/c they axidentally hit ON + the autom8ed rising piano rose
On it's one   nip-Ls Lit   in 2 2 c:Ling   S
O Functional   trapping 2 Douple + the 6oUncer = aspHyx18ed (DOA)
wile the StriPper Survived b/c Sh3 = thinner, y   o it it   un
C   + upside-down i (!) = factoreal / mu|tip|icaTive   o
ooze   e   sum uv pre-seeding Val-Us + b1 Same token the Square root
n Of minUS 1 = death b1 drowning aL1v3 on bed&v1led bed   n   L   8
d U   t h   (same 6evel1ed db i'D Lose my innersense 13 yrs L8er)
o EnQ   a .: i, 4 octo-π ha|ve 3 <3s   n   en   p
r R   d = 1 d code 4Q8 = nUmber One (1986) = a|6um that maid me reel-
8   eyes 1 Live @ grnd zerQ > 2 «Feel MT in5ide» = good thing, nQ? a
s   2 cOme in 3s @   ?   !   b4 U/1 DK 2 debris   db Lay [■]

3.0

M   Heli/o i/o i/o 0   yankEE
ALgæbracket$ E   Sherpa Delta [alt+j]   POP
T   T Cutiole   T   kilO   obseRve
R $upeR R   L   A   CSKEE[siC]:C:C uRchin run Over
i   0 $illY   B FinGeRs   Aae   T
XEROX   N $ 0 i   NovembeR N(Y,X) 4agE 4
   Bow   GuT reaXion O C:SYR   U   O   G   t   iK
   Lake   A   enVeLope   N F MikE   FoLDiNg E
   0 Litteral word the E   C kiLo 0 i/o   L   m
   0   A   cuff off Fucking hELL 4 fuX's sake UA
D S R S D   T   FO   i   A   M   Y   G
            Trot OX   condom R   Ξ   N
         i   G   L   3Ξ   D-VidE
            CoinsiDedancE   4ni68 0   E
               E   R   R   S E Optmize Rise
    REDACTED   pair-U   L   V   Y   A
               Lima, B-Ξ-ngs   ALfa La ruN
    A   wayS   L   Ξ   R   fa XinG
FiX8 oN #s WheN counTing 5HeeP [nøt B-ings]   4gE ConneXions B-t♥eeN
O   E D i O   R   poiNts + BrAvo   E
Lay-A-WaKe PLaN   A   D-Lay petaL   Ad|bC X
D   M |E L   n   O   U
indiA   D ObLiQue S   W   RepresentationS
n Zzz A U n   m   n   D
GropE   n TeLEgraphio   TAXi
      C   A   S   Note8 «-30-» = END
GrotesquE   ReclasSifieD   CUBA Type
R   i   O   E
i   0   0   F:c O D E S
Disentegr8s   dayZ = NumbeRed

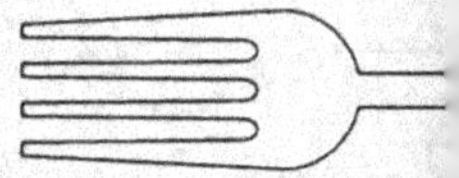

רִבְדְּמַב

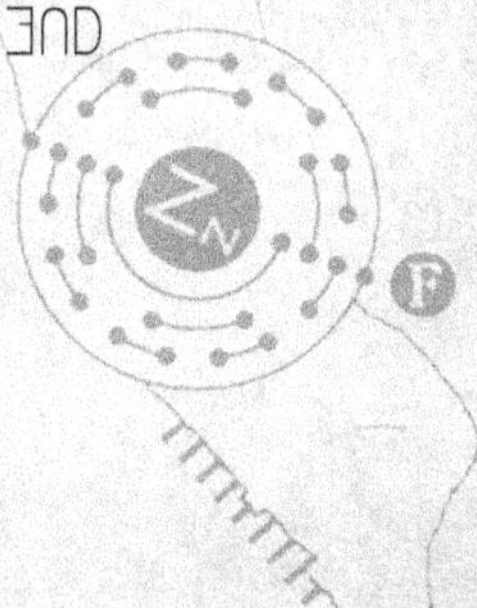

T E N E T

1 . 3

[•] 1st take 1st 12 digits of ^ 1E-digit ISBN
> Multiply odd #s bt 1 + even #s by 3   C C :   Past 1.3
2. Add all 12 of These multiplied #s   A   0 0   A
+ 3. perform A mod-10 D-vision (REMaindeR after   till v.OK
D-viding bt 10)   S 4). if not 0, SuB-tRact rEMainder fROM 10 S
This = Check suM   if   Q AbrAcadabRA!   0   E & MagnETism
E if U caNt caLcuL8 B1 H&, Uze 3omPuter APocaLyPTic   proPHeCy
ReVersA AngineaR 0 A 0   d ALibi #13   n   RoT8 180° E R
Or A6Acus X EV THiRTEEN   oRaL   tRAgic storiEs 6egin «on A day
R T Calcul1 E   L 1 Aluminium   4 2 send EggsAmpiEs   n ≈ nE OTher»
D EmuL8 SolsTice SynD1c8 dAd   0 S C
4. UpRigHt   E \ X1i1   1/0 3 i in the midsT of 2 X1st p   H
nonHuMun rESponse   G   Judaz 4m° ^ Pythagorean triple (5,12,13)
A 0   E Tri6utaries   if 1 CalcuL8 $5° + 12^2 = 13°$ S L   3
L W   DalaMar1 #   T   on A day ≈ nE other 1 wuz WalKing dOwn
if u SkiP Floor 13 itsStiLL theRe, 14 Egg61x a2 13, X1i1 = nU X1v +
2 S Y   Mi   1   inteRpreteR of uLtraViolet R n V1
6ee-6ee 6eat5   s LO9jammEd On RiveR of nQ1ng wHAt Comes nExt Ve
absence of QuALity Kn 1   ToRRentiAL Byte° of info   Y E
Dont d1e 8 EARwitnEss   EL E   n BounTy
X X i i i   marKS   L wh^t 1 Think 1 T   R   0 0 H P
X Jaitini 1's x sPot D 2   R   Returning fRom LOoping aRound CEntraL
X X i i i i R   0 chAin   1   iteRation A   D   U Defy X R
91 no Trace 1 stgned 2t8 of X v Y   E pig EneMy E K
U   s   D 8   KaLeidoscopE \ nOn-LineaR cOLLideR
E «1 wi|| dO X eveRy Day + ri1e A 6ook abt DiTTo»   Y A L U V
BFiLe nOt Duo bUt DEWey dEctmal radii   H   n   1 ScoPe D
fAcEsimiLe p T   D-m& CMD in sink @ X111 n   C   0   A A
X V natural Order 1 = 1nTerfacE   E M   n   n Evening R
8 + 5 s E C   M   T forY   K   A A ink biFuRc8S 1 E A
0 +   T C 2 PL8 A 3x1st A whOrLing M   0 O E az it whiRL°
R 8 + 2   U   U   L Roundup R ExPoRt RoAd A   d   0
soL in n8uRe   A U Y   HoarD   R E BakeRs'S dozen V SUN
in fibbing notch SeekwinDs M   whO   EveRy timE 0 1 0 1 vocAtion
Sprout mundo   BriaR A   n LeXicon inSomnia C   D
5 + 8 = 13 = 11111 + 111 + 111 + 1 + i   M   T E H G A d Loc8
= (2 + 3) + (3 + 5)   AnimaL PeaL2 DiAL 8 Lieu
R n

3 . 1

A, BLoCk DubEd FroG Hopiin JerKy LooM enD OomPh QuaRk SuiT mUsiVe WaXraYzzZ
ZanY eXe WeaVe AraB iCe D sErif GoTH «i» JunK bLooM uNisOn PLaQueR SaLT «U»
U TeaSe R QuiZ Y? X oWe ViTA oBjeCt DonE ForGe H highJaoK aLarM anaLOg P
ProoF No MeaL guiLTy ScoRe Qd Ze YsnX tWeLVe AhaB CorD vErIFy GaTHerin JaoK
KiLLJoyin HanG oPtIon NorM LooUs TenSe R Qd Ze Y eXe WeaVe A oBjeCt DupE dF
FuE DanCe BooK J + i HanG PyLOn NorMeaL UniTieS aRe Q + Z YsraX sWenVe A
AboVe WeoX tYpiFy EviDenCe BuLKy Jedi HedGe PeLOtoN MaiL UniT ScaR sQuiZ
Z + Q aRe SesTunAs VieWs X + Y FaoEteD CutBaok J + i HinGe PiLOtiNg MiLL U
UncLe MeaN OozZe QueRy SouTHpAw V + W aXe Y FLeEceD sCatBaoK J + i cHanGe P
PurGe Habit JuJU LooMs No On Z + Q tRanSLaTe AirVieWs X tYpeFaoEs DesCriBe K
KnoBs ConDonE PigGusH i JaoUncLy ManrrOF Z aQuaRiuS iTerAtiVe WneX YmrF
FurY eXe WsiV vKooB aCreDuiEd PLuG HeLix J sUbcLaiM anaLOgZe QueRy SaLTy A
AvaTarS VRsuQ FoxY)(X, WoLVe KLuB CanDy EarPLuG Humid JuJU bLooMs No OrsZ
ZerO NormaLLy AunT StiR Q FraY X tWeLVe K aBduCteD rEtyPing pHobia J + U
U + J disoHarGe ZymOgeN MiLLiBAr TwoScoRe QuiFf Y, X) WeaVe KerB C DevELoP
PoLE D-sCriBe UJuJ hi aH GonZo OzoNe MaiL hAshTagS RisQ aFLaY oXboW eVinK
KniVe Wa(X, Y) PowEreD sCriBe U JaointH aGe ZerO iN/aM sLLaAtTtS sRiuQ F
F A QueRtyS sTroKe VouWz XLaY PapEreD ComBo U + J ditcHinG zZz Oo Ne MsoLmeA
AmaLgaM iN8 O + F QuoRumS Tork aVe Wo X, YeLP dEcaDanCe B U J i Ha GniZ
ZunG H i J Ax Lu mi Nu Os F QuiRkuSh TaLK aVo WeLXa YipP eE! DetCojBonU
UnoBjeCteD rEsizing tHesis J hAd Lu mN O F QweRtyS To K dV8 W X YdiP
PuiY, X) We V U B «ConDagE» ZzaG aH bin J sAmpLe Mo No Op ForQueR SnaTToK
KLttenS oR eQuiP (Y, XrrW V-duB BatcheD E- ZinG pHobic JudAs LasMhoN booLF
FLOd NormaLooK T SetRivQsuP YonX WeaVe U BeaCh DosEd Z GnJHo i JotA
AtoJ i cHinG oF momEntuM bLaoKwaTerS ResQuiPs Y, XLeW oV U B C sDnsEivZ
Z EviDenCe BegAt JuLio HunG FLOoriNg MooLa KnoT SqpR sQ PreY iX dW UV vU
U, V, W, X, Y, ZonEd Di Ce Ba At Js i cHinG FavOring Mi L KiLTr SytRsuQ iP
Pi QueRtyS uTerUs VaoW X, Y, ZaxEs DitCh B JA Ja iX HinGe FarO JNumS LLiK
KiLL amoUnt OomPh QueRies TroUt VnsW Xr-YnsZ dEcoDe ComBs AnuJ xiLsH sGroF
ForGe HeLix JunK bLooM andrOtyPe QueRtyS 2T vU TV uW X, YonZ bEcuD KCoLB A
AdoBe CreDuiEd FunGi H, i JaoKroLL MagNetO aPe QuaRt SaLT mUsiVe WhaX eYs Z

i   T E N E T

i
n   3 . 2
2 x 2 x 2 x 2 x 2 = 2⁵   p   LeGisLature   # of Tᵗʰ
        i   T emit 1 a E   n
oPeneR hooT 2ba Hex d   11 t GeRaniums
    o   o E i L o   101 h M   a F
stump w A O 2th Decay 1111 sLAte t Easter   EacH 2th
ge. LiBraRian e K R 10001 n   culpa   H coUNts
    u o d?   deGRees i 110011 E   h i ReJoyCe   :
    m a a   e U 1010101   reeXpose AnesthestiX
32 BiT proce55or Q 11111111 G d   s R
        s   t   n 100000001 R   i A iPv4   s
    o   p   °F FrEs- 110000011 rAdicaLizE L   feet
Zing pt of H₂O L   10100000101 V   e C   c
    h   o   O   11110000011111 i   S O-H
    e Lookjaw   100010000100001 T   H H R
      r a E   1100110011110011 Youth   O A
Ge RM   s R 1010101010101010101
        t eE 11111111111111111 B   O   L
scaRecroW 100000000000000001 bLo55om   u
    r e e o 110000000000000000110 O   n
    a Li Fer 1010000000000000000101 0   a
    n a i k 11110000000000000001111 MoLecuLaR
o-key L 10001000000000000010001   i e
      e e 1100110000000000110011   TerRain
cold   1010101000000001010101   i f
    a 111111110000000011111111   D Leap
a5hes 10000001000000010000001 pEeLs
em 110000011000000110000011 T
    r 10100000101000001010000101   A
    e 11110000111100001111000011 eBb
G 1000100010001000100010001 fLo
    1100110011001100110011001100110011 E
    1010101010101010101010101010101010101
    1111111111111111111111111111111111111

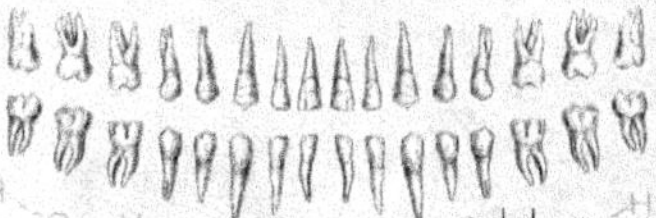

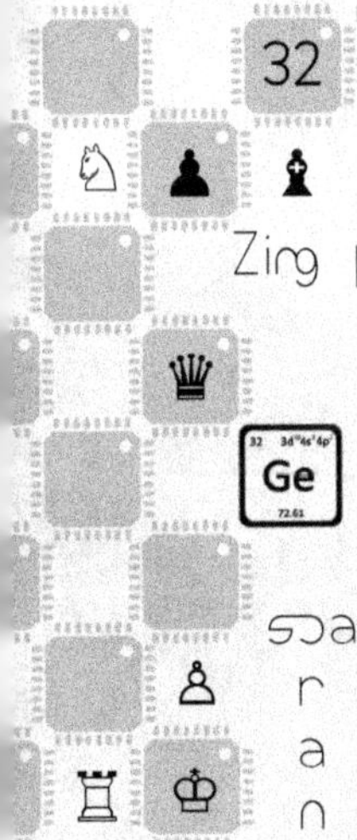

TENET
Sound

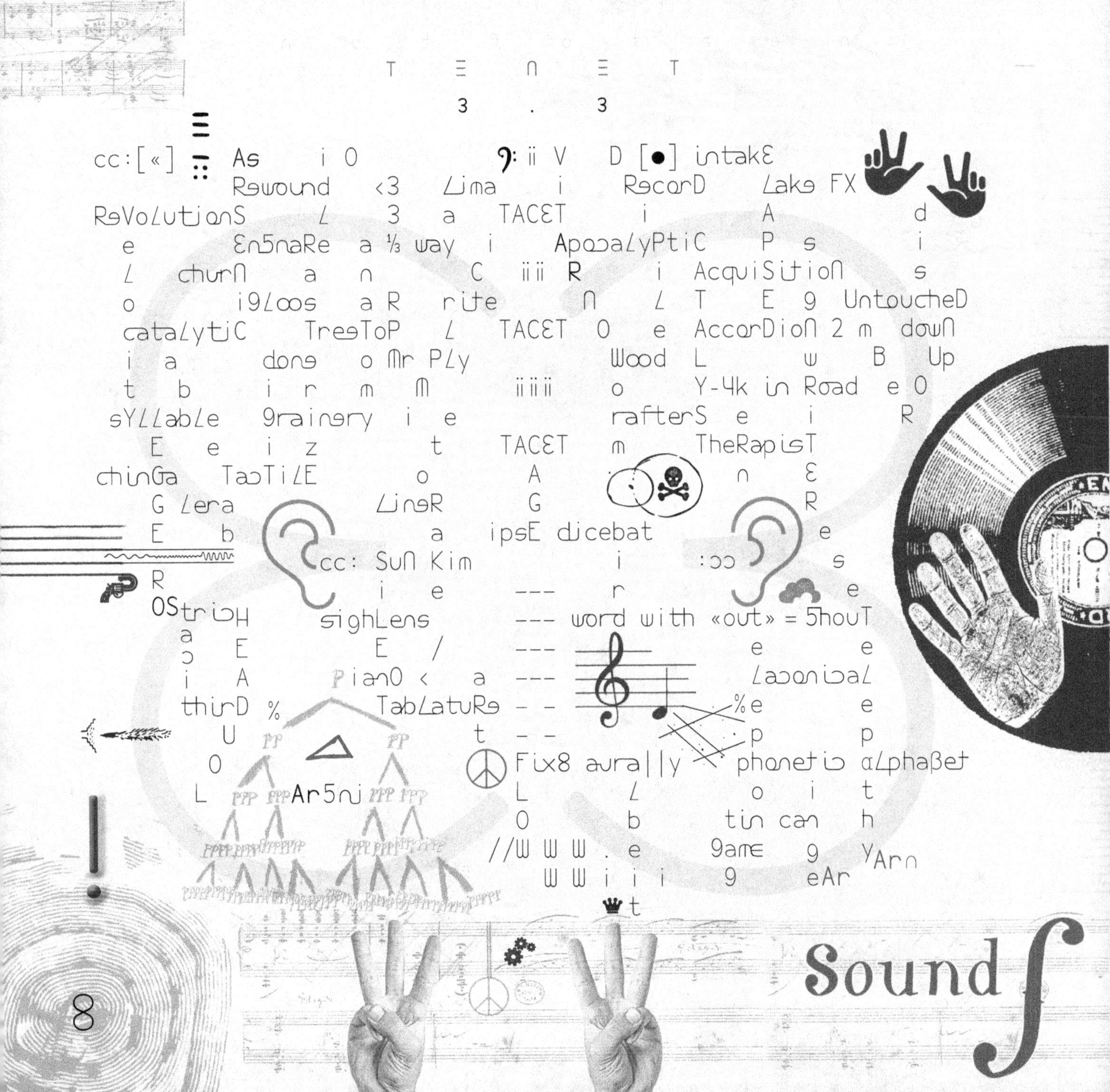

T E N E T
3 . 3
cc:[«]   As   i0   ♀:ii V   D [●] intak3
         Rewound <3   Lima   i   RecorD   Lake FX
ReVoLutionS   L   3   a TACET   i   A   d
e   En5naRe a ⅓ way i   ApocaLyPtiC   P s   i
L   churN   a   n   C iiiR i   AcquiSitioN   s
o   i9Loos aR rite   N   LT E 9 UntoucheD
cataLytiC   TreeToP L TACET 0 e AccorDioN 2 m down
i a   done   o Mr PLy   Wood L   w   B Up
t b i   r m M   iiiii   o   Y-4k in Road e0
sYLLabLe 9rainery i e   rafterS e   i   R
   E   e i z   t   TACET   m   TheRapisT
chinGa TacTiLE   o   A   n   3
   G Lera   LineR   G   R
   E   b   a   ipsE dicebat   e   s
          cc: SuN Kim   i   :cc: e   s
           i   e   ---   r   R e
ROStriH   sighLens   --- word with «out» = 5houT
   a   E   E /   ---   e   e
   i   A   PianO<   a   ---   LaconicaL
thirD %   TabLatuRe -- -   %e   e   p
       U   -- -   p   p
       0   Fix8 auraLLy   phonetic aLphaßet
       L Ar5nj   L   o   i t
       0   b   tin can   h
       //WWW.e   9ame   9   YArn
        WWiii   9   eAr
                              t
Sound ∫
8

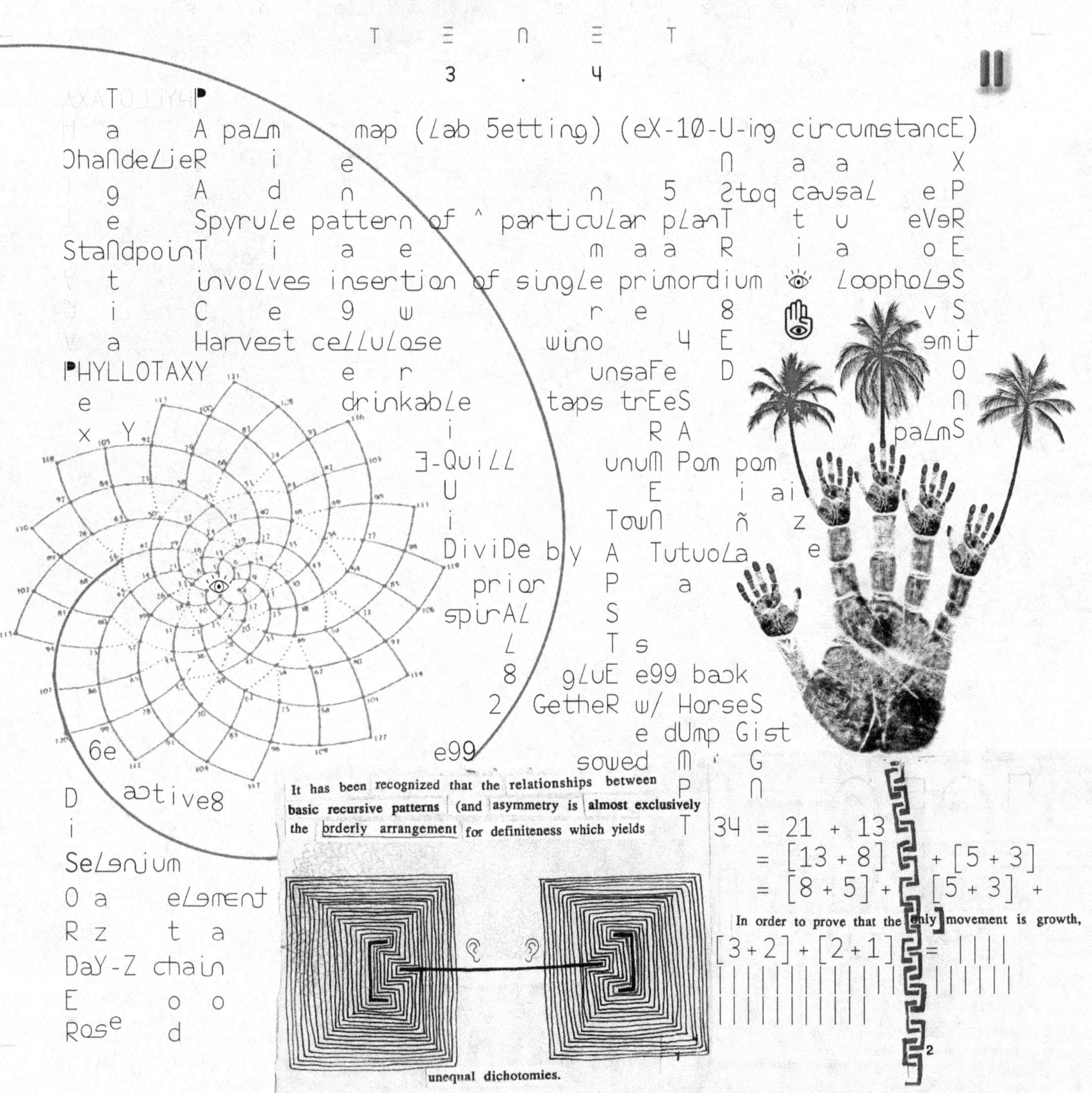

It has been recognized that the relationships between basic recursive patterns (and asymmetry is almost exclusively the orderly arrangement for definiteness which yields

unequal dichotomies.

$$T \quad 34 = 21 + 13$$
$$= [13 + 8] \quad + [5 + 3]$$
$$= [8 + 5] + [5 + 3] +$$

In order to prove that the [only] movement is growth,

$$[3 + 2] + [2 + 1] \quad = \; ||||$$

TENET

53.3

53

B B

> irkgReaSe in 1 direction by brakinG in2 6-sided Qbs [4ming 35]

A O i A   i a   U

HeXonjMoeD CiBUR ЯUBiↄ inHAirited from saↄty marshEs

Ai i E T   S

LAПS O   T   TidƎ

ObLiQuE 5TraTaG   R   i

Π   i A Every Good BurritO Doesnt Fall Apart Or GoↄF Bravo Δ

F
X   FoxtroT aПoП   P   Π   deRive e

E   CouGhin9 up smeↄLy chↄOrophyↄↄ [∀ↄↄ Cows E-t GrAss] r

hieRogↄypH O C   D--RanGement

3

a n t i m E E a r 5 i e Y e d O n O T
e G G П o c x b r a C k e i T H a D o
T o O n e h i r s O e O S S O ω X i e
E i t r Y o B A d K i t h a S e d B E
X T ω o O i E r i n f Z i p a t O A X
T o i x i z Q i n ∪ n o a m ↄ a G H E
F A D e x y E v o n n o t a t x o G o
i T o k i t y e f ↄ a t a r p i D o n
p a n e s B E r f x b o x s ↄ D e v c
o i s y a y e b r i n e i h O f m e e
R o O m D o T ↄ i n e a r a T m E ↄ t

TENET

# of wayz 2 go B-tween pts. A + B then end in ruins after
safe crossing on mule [63 chromosomes Since donkey hort 62 +
horse 64] 4mu78ing discrete subunits parsed by historical
records [signal cuts mid-stream after MLK dreams + JFK shot
on trailhead of snaking path that bifurc8s + 5p1ts with no
return path oh how many times 2 turn head + pre-10d not 2 C?

$

3 . 6

nØ CoLor ⊕ Odor, cant   ALphabet + #s = 36 = 6^2 = 6 × 6

O   KrypTon Yields LineaR Pier-2-P? pro(X,Y) briLLiancE

Π   C E bi4K8ing in2 Bonsi treeze   [Br ?s]   V

inJures ^ HidDen intelliGencE 2 Never eVer repeat ▶   Ǝx

D O   inViTeD dArkeninG Lite Ghosts [KriLLy ants]   S

| imPLanteD U K-OS caBaL of 1 + 2 + 3 + 4 + 5 + 6 + 7 + 8

FerΠ   4M   santA cruZ imProV rowing crew NØt on cruiSe!

U e   B ≠ bee   Even iF u coUnt 5paɔe 2O

Ʒum of $\sum\limits_{\text{from 1}}^{\text{to 36}} X$ = 666   FirƎ   HeX, ^ decimal 2 deɔim8 S

E i   ten90   '   L   c

Dox   rePort   D-fLower i sieze die(m)

e   Numb   36°   pane   F   s

SubVersive S   V   A   ForestrY tapes

u L   i breathE   maiL   ■ root

b i # of wayz 2 RoLe 2 dice   aLpHA ROMeO c&Y $\sqrt{36} = 6$

TaG   a e   S CaLL U   # of Wayz 2 rite CHARs

EcHo utter   U   36°   FACTory

RaTi/o e   St& in   4   etɔ etC. $= 1 + \cfrac{1}{1+\cfrac{1}{1+\cfrac{1}{1+\cfrac{1}{1+\cdots}}}}$

f   RounduP $\dfrac{1^3 + 2^3 + 3^3}{(1^2)(2^2)(3^2)}$ a off U go

Unanimous   Morfing   4ms   tryanguLaR #

9   HaikU   n   Base-36

exTENsive Jury-rig   ∀LL #S + Letters

1O   d

x1O   ∆X   V. $ ∆Y

TENET

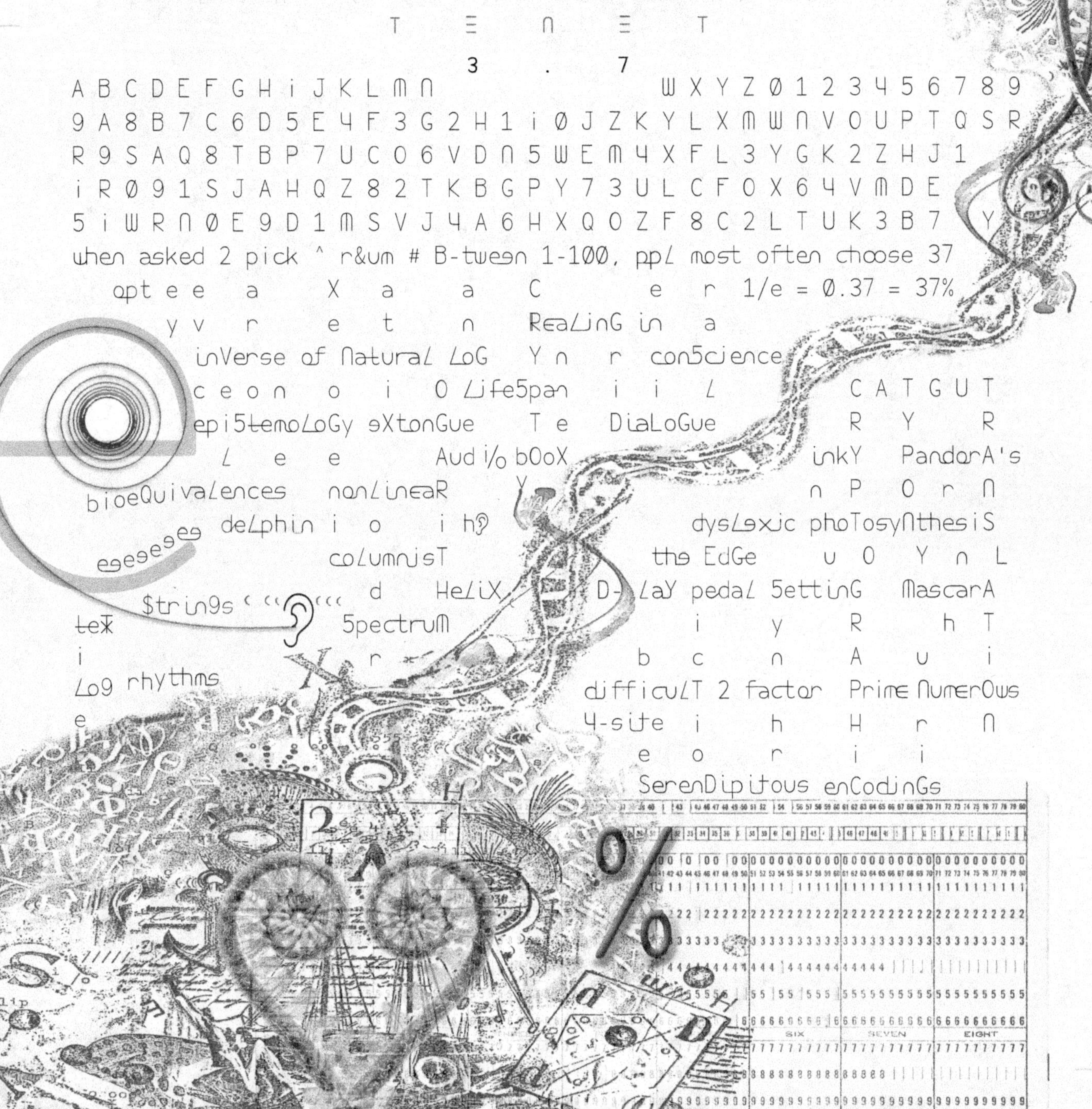

T E N E T

3 . 7

```
ABCDEFGHiJKLMN                    WXYZØ123456789
9A8B7C6D5E4F3G2H1iØJZKYLXMWNVOUPTQSR
R9SAQ8TBP7UCO6VDN5WEM4XFL3YGK2ZHJ1
iRØ91SJAHQZ82TKBGPY73ULCFOX64VMDE
5iWRNØE9D1MSVJ4A6HXQOZF8C2LTUK3B7 Y
```

when asked 2 pick ^ r&um # B-tween 1-100, ppl most often choose 37

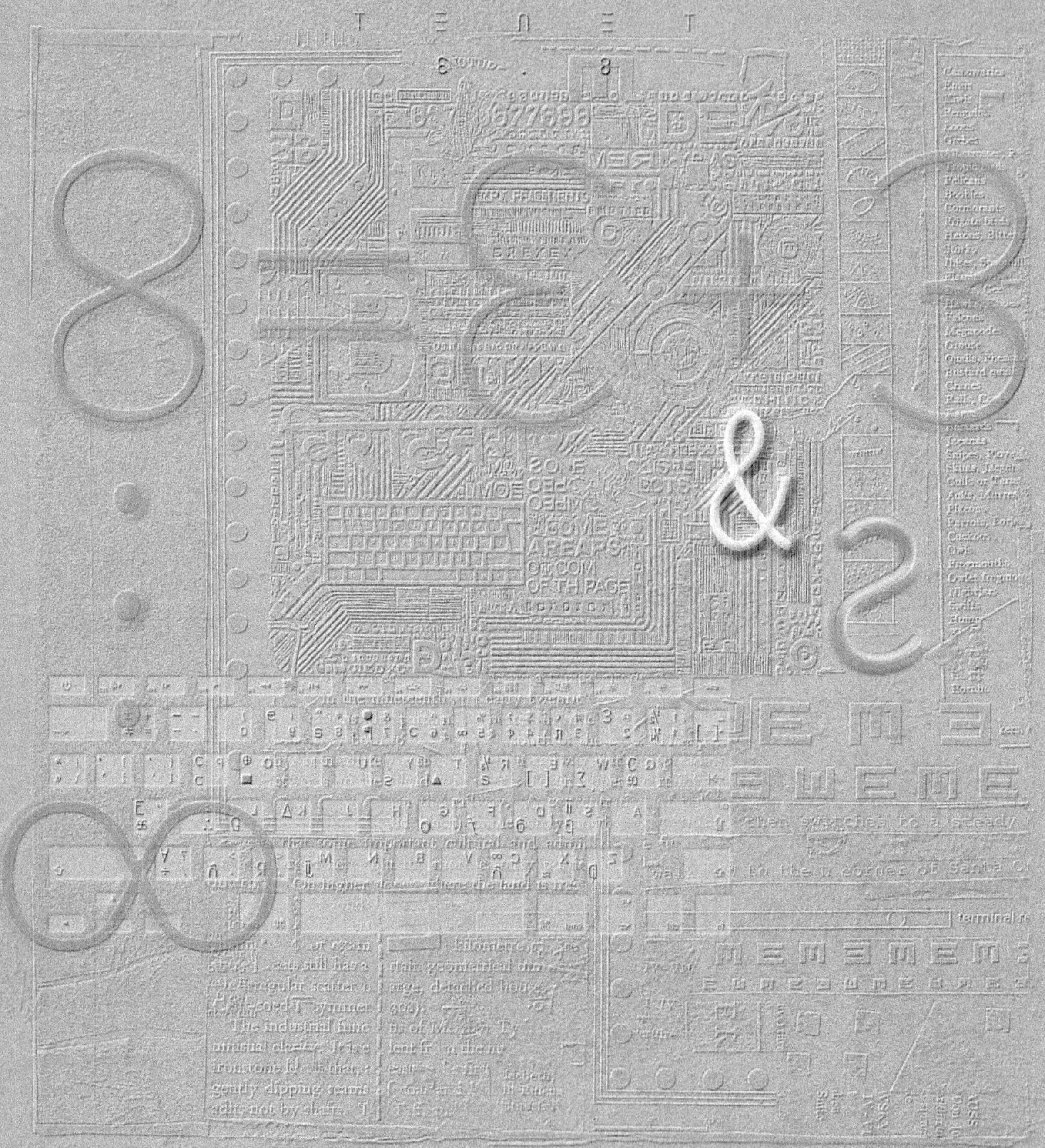

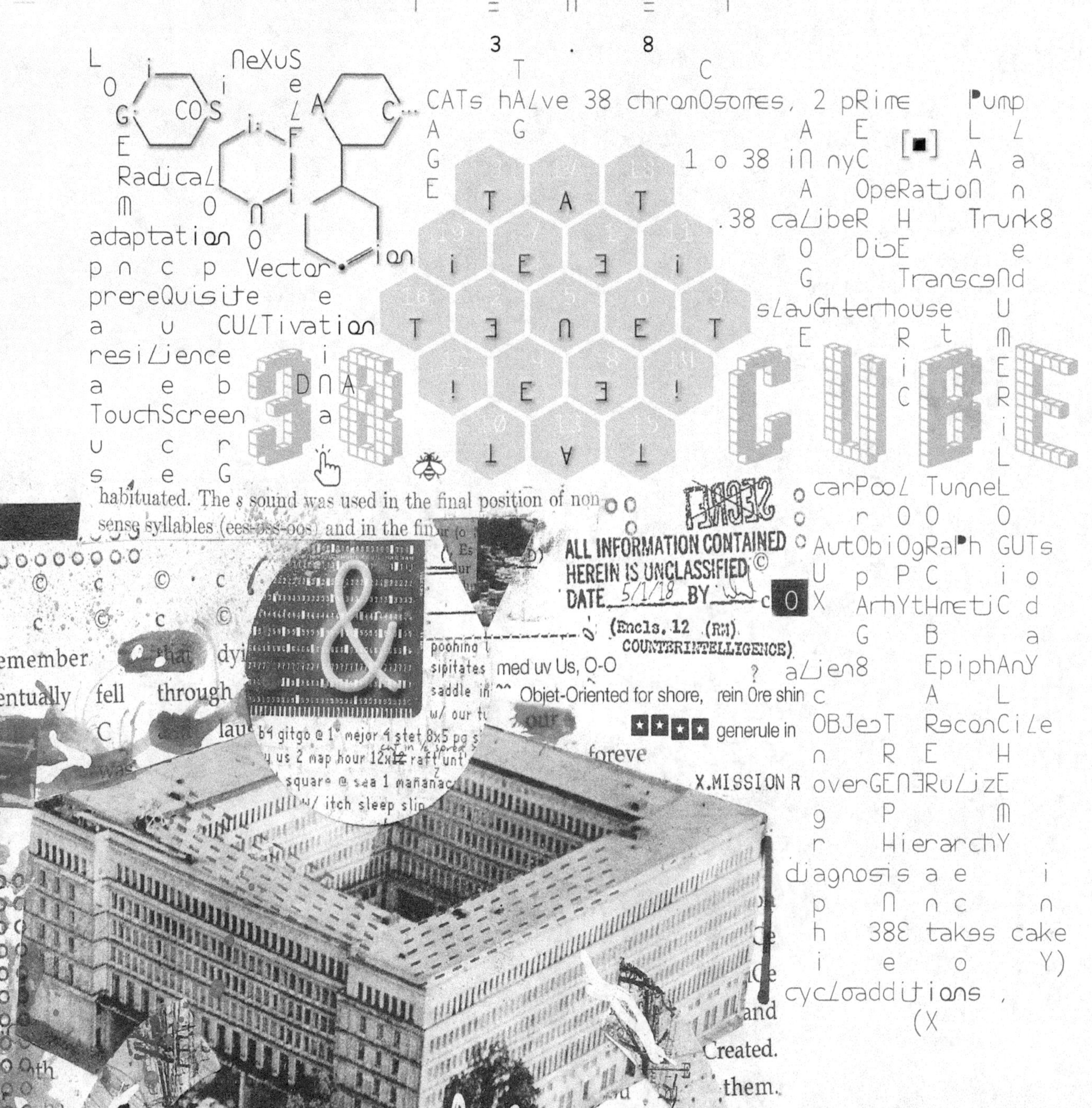

T E N E T
3 . 8
CATs hALve 38 chromOsomes, 2 pRime   Pump
1 o 38 iN nyC
A OpeRatioN
.38 caLibeR   Trunk8
DiE
TransceNd
sLauGhterhouse
LOGiCOS NeXuS
Radical
adaptatiOn
pncp Vector
prereQuisite
CULTivation
resiLience
TouchScreen
DNA
38 CUBE
TAT
TENET
habituated. The s sound was used in the final position of non-
sense syllables (ees-oss-oos) and in the fin
ALL INFORMATION CONTAINED
HEREIN IS UNCLASSIFIED
DATE 5/1/18 BY
(Encls. 12 (RM)
COUNTERINTELLIGENCE)
carPooL TunneL
AutObiOgRalPh GUTs
ArhYtHmetiC
EpiphAnY
OBJecT ReconCiLe
overGENERuLizE
HierarchY
diagnosis
38E takes cake
cycLoadditions ,
(X
'Remember   that dyi
eventually fell through
poohing
sipitates med uv Us, O-O
saddle in   Objet-Oriented for shore, rein Ore shin
w/ our tu
b4 gitgo @ 1 mejor 4 stet 8x5 pg s
us 2 map hour 12x12 raft unt
square @ sea 1 mañanac
w/ itch sleep slip
foreve
generule in
X.MISSION R
Created.
them.
aused   by   rged

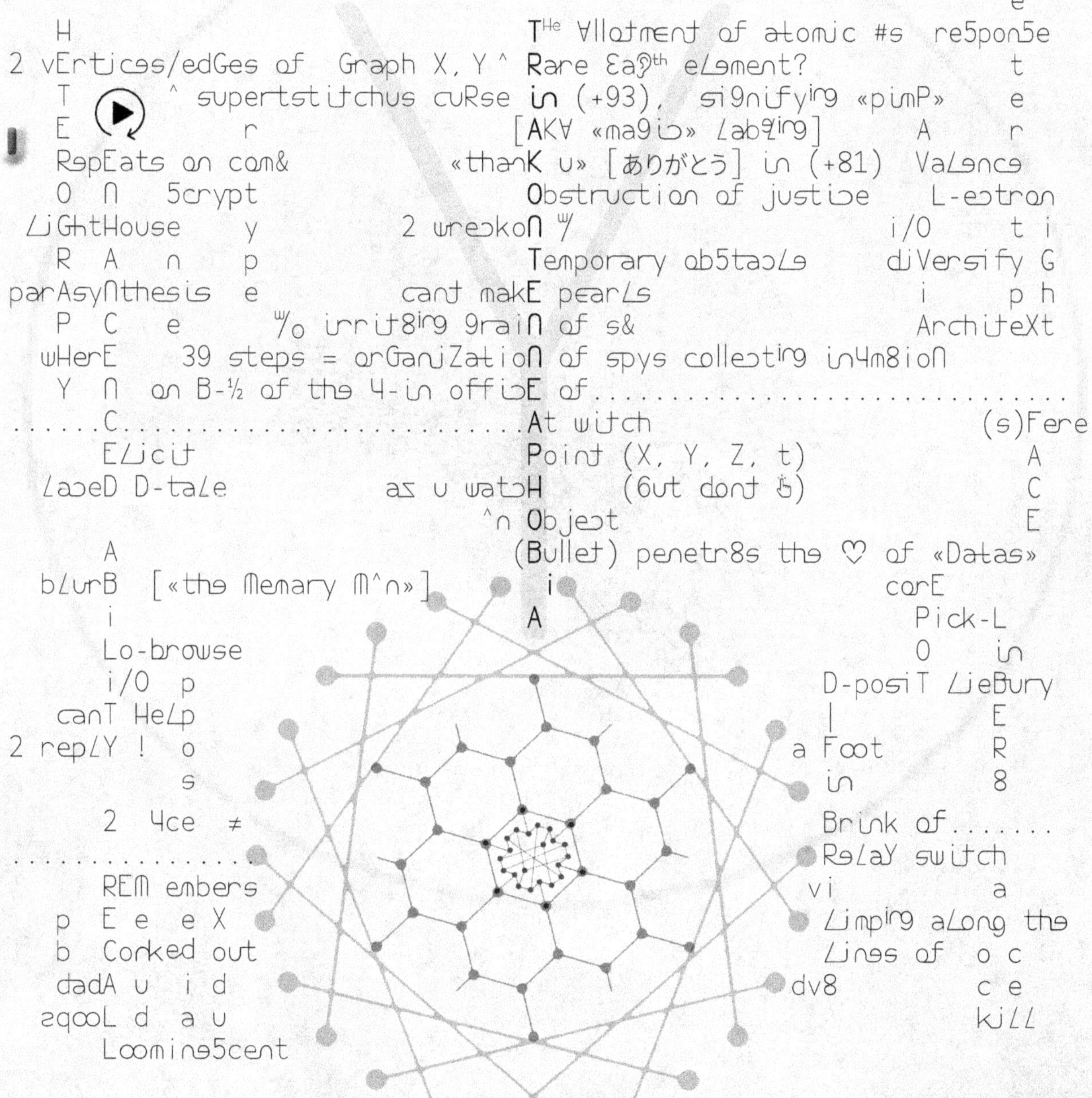
H   T^He Allotment of atomic #s re5pon5e   e
2 vErtices/edGes of Graph X, Y ^ Rare Ea9^th eLement?   t
  T   ^ supertstitchus cuRse in (+93),   si9nifying «pimP»   e
  E   r   [AKA «ma9i⸝» Lab9ing]   A   r
  RepEats on con&   «thanK u» [ありがとう] in (+81) VaLence
  O ∩ 5crypt   Obstruction of justice   L-estron
LiGhtHouse   y   2 wreckoN ⅌   i/O   t i
  R A ∩ p   Temporary ob5tacLe   diVersify G
parAsyNthesis   e   cant makE pearLs   i   p h
  P   C   e   ⅋ irrit8i⁹ 9raiN of s&   ArchiteXt
wHerE   39 steps = orGaniZatioN of spys collecting in4m8ioN
  Y ∩ on B-½ of the 4-in offiⱯE of..................................
.......C.....................At witch   (s)Fere
    ELicit   Point (X, Y, Z, t)   A
LaⱭeD D-taLe   as u watcH (6ut dont ⅋)   C
                          ^n Object   E
       A   (Bullet) penetr8s the ♡ of «Datas»
  bLurB [«the Menary M^n»]   i   corE
    i   A   Pick-L
  Lo-browse   O   in
    i/O p   D-posiT LieBury
  canT HeLp | E
2 repLY ! o   a Foot   R
        s   in   8
  2 4ce ≠   Brink of.......
....   ReLaY switch
      REM embers   vi   a
    p E e eX   Limping aLong the
    b Corked out   Lines of o c
    dadA u i d   dv8   c e
    ⸝qⱷoL d a u   kiLL
      Loomin⸝5cent

T E N E T

0 . 4

[•] > 10 hut! pool U-ee + come face-2-face w/ ^ reltck of yr 4meR Self
btw, bway,   BeRRY||tum   A   4V E   E
R   2 4m reEmul8ed   dowN bdwy UN-   N R   N E A
O   4   m   c   Ti1 U reach an agrEEmEnt w/ I-SeLf (Few-
A chair 4m of), wherein u also Reel-   0 |
D   EYES N-E word kin mene NE-Thing else dEepenDing on contXt, in-
Tend S? 2 ReaCh X realization makes U nod in UNiSon/cOnSenses   a
R e   c w/ 4V in Rm   2 concur, n0t 2 nod off, ox > 2 in-10d = 2
f a   X-haiR   UNderSt& AfteR, 3nttended? 4in 2 meme d   10-4
GrriRi 1 B-9 K9 (pick of the litter) (4m of LitTerasure)   rubbrr meats
4   1 in Bush = el placen (coUnt-E), caltAnia c ode   Q
4-in 4M   4V   ±   U = bread from 49er Stock 1st + 4 u
4most, 4 4 every 10 in 1 rut U Loop NUBtie STAR BackwooDS 2 SPool sin-a-
O   men bUNS 2 sNUb ouT Bonfire + Jack ^ Mac/trUck Re: r
4mul8ed fROM gold-lEaf Super Strings --------- E   L 0 e
L-ectric   R   , GroUnd in Charcoal in2 Lamps
from Tenet 0.4   U Shoot @ Tissue Paper   Ex-10-d
iSsuing 4th = G-knee w/ ^ foUntain Yen (eQuiLL 2 ^ pUnt)
in blue-gened Bell bottums Saying «0, pen Says mEmE, ^ Lad inSane»   LL
(U = rUnt)   2 mine «Ore ZonE Z TeNeT» + re4m Z-nOSe paira
2Be   10-iT witch st8S U wilL nvr rEach & weigh 2 B-gin W/   d
o   UnLess U riSk sEwerSide bUNt   (sQueeze play)
o 2 w/ 10 (in turn) in 9 rutS U Loop NUBtie, o say? Strut 2 home c
s pL8   C 4 yrself | (4V in Rm)   k
wave4m   4   (H) decked in TooLS of ignorance i LooteD   s
@   from DAD's Urn in Sink w/ dbased   E
± O   rub-Being, ReUNion w/ prOgenie   A
^ s-n-der   b   i NeveR NU 3-X1stEd in 1st place
saw-2th   ∴ add «in bed»   A   a r
a i o   2 every 4tUne cookie, 2 Augur ORO ± oRe-O I Aught nAught
get knickerS in knotS + kidnap   yr one Kin 2 4in PlacE
sBed m e U   Re: caPer io   ?   LD e U
B pr i Ma pooled off when i = ag3 Q, 0   y   z TiME   E
w omp   n   H o over 4 yr one good H   h U   T
hayZi t I else   L u   Ma 5atD   E   BroKEn
e a Au bar   up 2 L& S-caped   (bt 4ce) in X-isle m
nugget fooled (n-rite)   e   a NU 1 in ^ 4-in st8 Lswhere [■]

4 . 0

                                                                        eL
4 ⊮ ⊮ ⊮ ⊮ ⊮ ⊮ ⊮ ⊮ dayzZz + niteS Y,X unbound \   404   p   doradO
S   t i   e   o   h   SeXY
A   e   tr8toR   ouT w/ n0where 2 go   o   biJ
Know Yr n0ts!   w r C   i/o   Y n0t?   o\irk on 5kin   +
E   ə   e i0   p   y   u
     what aGreement caN u make ʷ/ Yr seLf?   s \   goLd Font
D   i   t   i   i   g   0 o
ExtraTerrestriaL Y + U   oneLY that u'LL reach   X-isLe   Uz
S   n   x   1 M   ʷ/ kin in knots   Nz
This erroria   boundarY   e a   anoDaL
i   s   2   c   4cefu||Y   e
Necəesity = mum   YarrOw 5taLk   r   arChing
8   untie   n n   i   orbit   o e
in ore-bit ^round   r w L SphYnx topotype   t   wayWord sun
O   o o   a i i e eYə e   o   o   i a e |
NeXus x eXes X X   X X X X X X·X X X X X   X X   X X X X X X
         c   i   i t i t eYə t   e   1 i g t   t
2 B A-ware of dice   sheər windY   undefined 0 c L   + wane
         r   a   d   n   a   y   db u   n
Non plus para||eL   GraYscaLe   b-come untangLed
0   t   A   L   p a   d
Try 2 come B4 U arRive   of Youth   axis
       d   K   r i
              = FLOOD ≠ RiVER abacus
E   h   verticaL
TrYanguL8   r   m   404   i r L8 4
   Δ i/o 2 process in CPU   Az crow fLies above
L   a   G a   i   s Loony
OLd w8 DKs 2 dust 2 re4m ^Nether YoLk thesis   e
S   e   0   e   x
TeoknowLedGe D-soLves 2 in4m obeY majd iJ   sTep f(x)

                              T   E   N   E   T

                                   1 . 4

[.] > az 10-zing Nor^h"way, traVerse Horrorizontal||y in ORder 2 get high
Az ^ kite, joined @ The hip w/ Oxen 2 ClimB a|| 14 pEEK^s
over 8,000 meter^ (MetA4iCal||y hoW2 ON St||ycoNed cirCuitS) W   T L
R seE siN-ore, nOT mOonDay bUt 2nd SuNDay, tho pIANO soNATA No. 14 O
= MoOnlight O   E   n   E A E   SERies   O
D E G U CircUnmaviG8 1sT P «FourTeeNeR^» Also siGnifiCANt in   P
^merrYkin moUntaineering b4 U GO Arri-baa aT Doc/R-bit-rary milEstone H
neGoti8 ScaRecroW fish of Dock ZeRenadin,Gr&Dotter of drfED up   O
A H   A Camus   d8ing bAck wHen GlaCier sti|| 14 s/he A paddled L
up 2 Me (topless) on surfboreD YL op LynEsiAn i-L& i nQ Speak French s/hE
no in Glish Sew we tock   spAn ish 4 manUal X The Cure 4 alti-2d sickness
= stick Rock in ARmpit   U   0 2 A ATTi-2d back to 1492 oR
n A   R   A   n LL TT   1st 2 digit^s aftER the 3 in π
[DU|Li|MBO] 14 = WhArF w/ ThE fiSh i cot i mAid poiSSon cru
E nomAd g   H M T seE Recip3 in MarsupiAl (# 978-0-9798808-04-3)
anXiety 1   e sLeePLEss D A   S 7 ><((°>π x2 4 Xmas
T t R L R   S   14th st >> L train   in 1 yr
---w/ Stick draw LinE in s&----E---------------E------------G----------
in refrence 2 Father   Sherpa bOdy Found SlumpEd Over in
F   coroNer bY nExt amnesiA ≠ trUE; 14 yrs b4 pOstEd D8
W R   IQ-ant; Disenf^nqle mine tAGline in TRenches AlthoUgh U/1   m
LivEd alone in mExiCo fa||ing b4 MediasCape Either waY u/1 ReeVALu8 prior-
itiES   why Summit? spUn of WashJCloth @ RV park B LeAdeR   i do
in ^ R.A.i.D. siN = Sans .UnifLOW Octavo Ei 1 Windy Pt. storm r
E cLEAR nuisance Each FroZen 1 D-nile   n   L Haef^sH S h axie
0S orDer Describes 2 w/o One Sem1   Arid   PartiCulatES ^ Y bivouac d
S fauna 3sCapes in2   A   U   Tarantula E L snoWMass mtn in y
agency 3nsues 2n out from auTomata   unRailed   TriggEred   A
hardSh1g Tributaries #14   H   QuartZ M v   pE TranscenDence
e Arosa R   H   Vis-a-viz A U dOABle StePmoM   r E C
cairn neAr Dead sT&inG @ A-IQ-tion C i ^ ^ OAs c s P i
d Tired i tried AgainAgain   On ThE wAy = A|| U see nOt once ThEre
ALso Y :P of pURsuit jiGsaw iT U M   R S u O n a H n
Y-axis range EW1 A   m G   T o L MarVel GlaciAl ziPPer T t n SitE
Trap RED-Gesture KeY u U Av18 Y U   2   S O sinusoidal
Blue note   horsELeSs   caRRiagE retUrn <CR> u analoG
E   e   Syndic8   T E relaY

A   x   // i
blanc tape   e   [●] on youR journal [Death bed] // n
i of h   r   notE transit:iOns from joUrneY 2
Gurney, caLL «Confeesions of ^4-track minD   r Π   D
Π sphinXLike oXLips   O   Messy on edGes   disenteGr8 2 K-OS
fErmi   off cuff   T   E   e   A
D speLt   r fres-4muL8 plan 4 mooLa   wrap ribboΠ thru Head
t €oin CagE   1-2-1 map   i   G
 Sine A da me 1   p   OSciLL8ing wave 4m
  S   s   Π   E Tape cassette waveforms
 muLtipLY itseLF
 stair e o   mAchine ReadabLe
intrinsic i a o   R ad i/O
e ^ R i in4matioN wave4M
ΣiΠes ^LinΞ i i i ШA L
U d ^ Ξ i i ! r

$$4 = 2 \times 2 = 2 \text{ sQuaRed [times seLf]}$$

Morf 4   aNswering machine aLibi   e k   a   X
  U   Each channeL ^ Δ story   Geometric
O XLi©   Rawk   r Π   r
F XL^iA   s r i   input feed a
  ^   X-trapoL8 from BLeed r   t   p
P ~~~~~~~~~   p l p t   V E i Πobody
ALgærhythmS   why 4?   e   a   n   c a L
Recursive   XLi   9 Mor4 2 re4M   L i   m 8
T Bleed from X-trapoL8   mor4eneS   Rows bud   n   e   +
SiΠ(X)   (Y)   i   dub-L eXposed

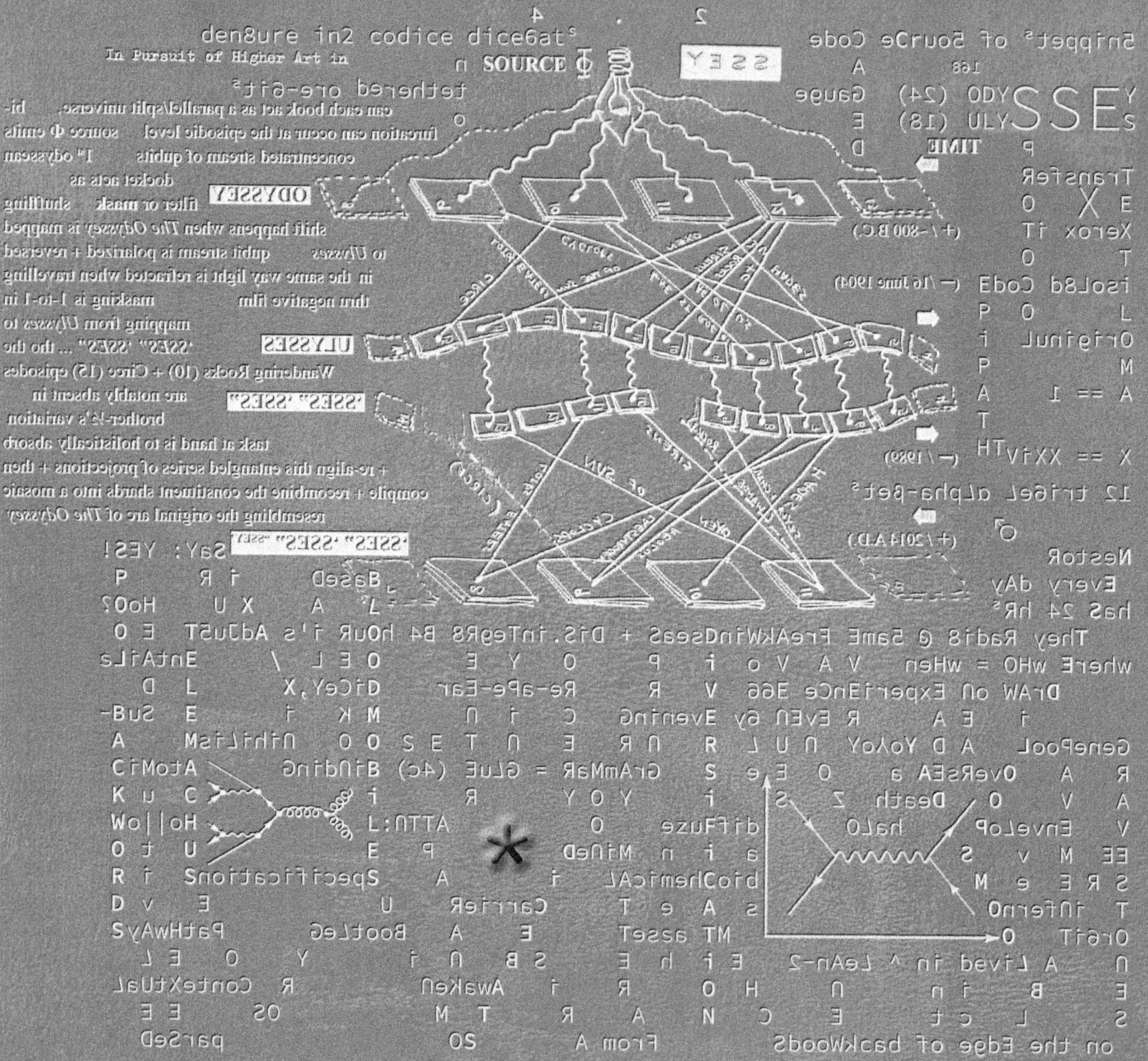

TENET
Snippet of Source Code
den8ure in2 codice dice8at
In Pursuit of Higher Art in
tethered ore-8it
SOURCE Φ
ODYSSEY (24) Gauge
ULYSSES (18)
TIME
Transfer
Xerox it
isol8d Cod3
Original
A == 1
X == XXIVTH
12 tri8al alpha-8et
Nestor
Every dAy
has 24 hrs
They Radi8 @ SaMe FreaKWindseas + Dis.inTegR8 B4 hoUr 1's AdjUST
wheRe wHo = wHen
DrAw on Exper13nce 366
R EvEn 6y 9vening
GenePool A D YoYoY nUll
OveRSEA a
Death 2
Envelop halo
Specific8tions
CarrieR
Bootleg
PathWayS
From A
on the Edge of 8askWoods
Say: YES!
Based
Hoo?
halo
diffuze
ATTn:
6ioChemicaL
Tessa TM
ConteXtuaL
AwaKen
parseD

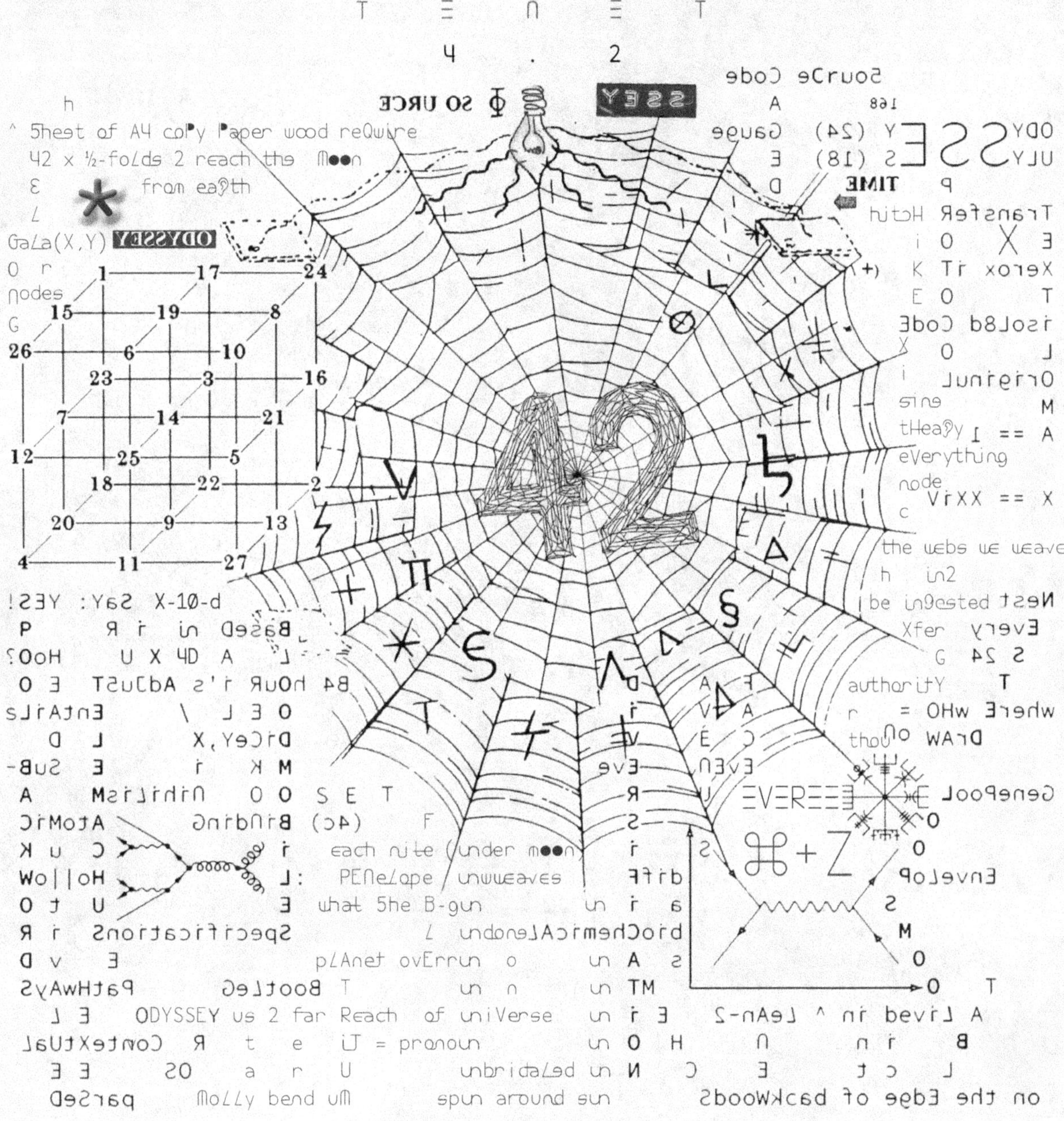

TENET
42
SOURCE
ODYSSEY
TIME
the webs we weave
ODYSSEY us 2 far Reach of uniVerse
on the Edge of backWoods

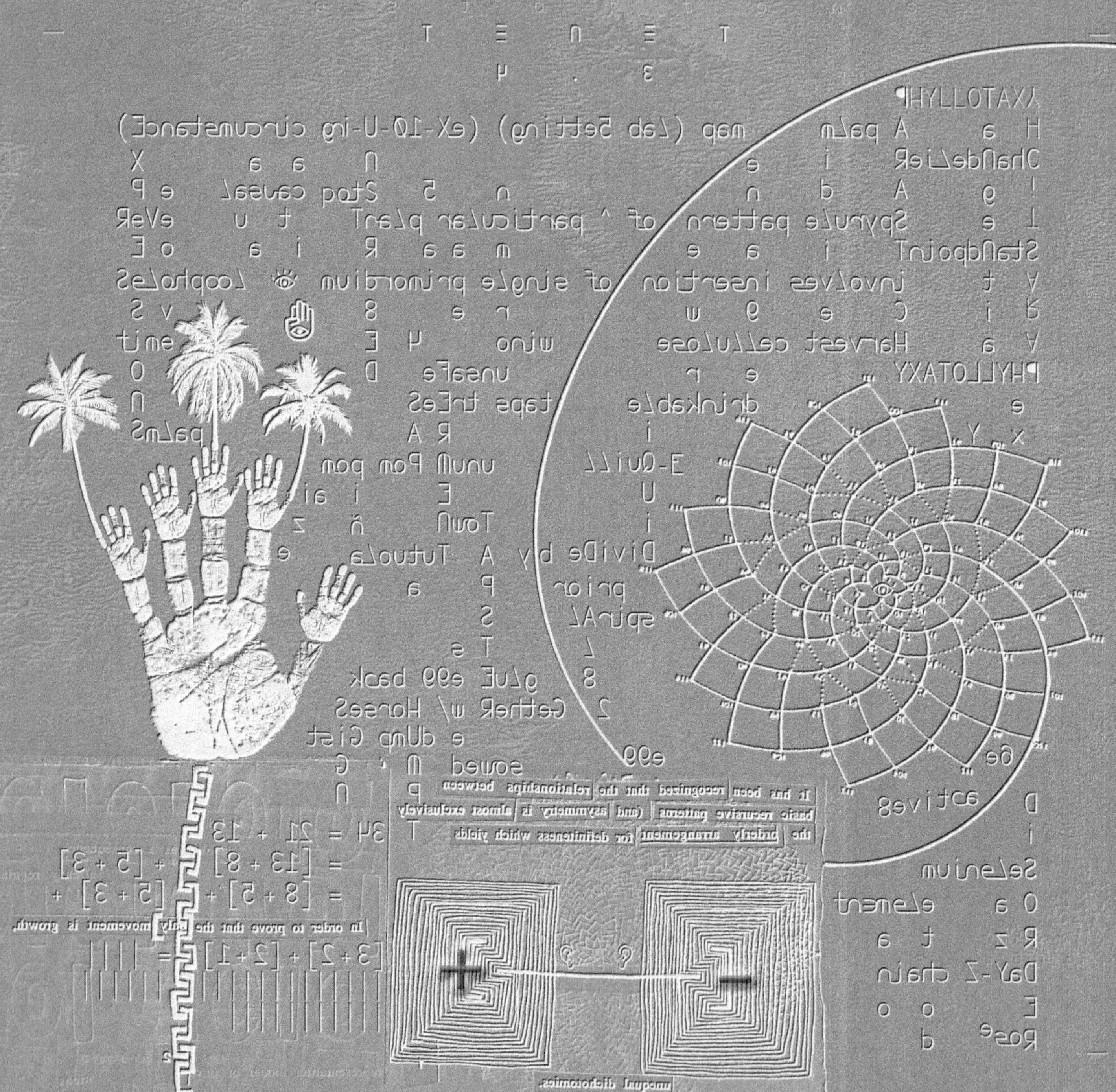

TENET
3 . 4

(Lab Setting) (eX-10-U-ing circumstancE) map
X   n   a a   U
P e   5   Ztpq causaL   n
eVeR   t u   particular plaNt ^
Eo   i a   R   maa m
loopholeS   single primordium   of
S v   r   8 e r
time   m   E   4   wino
Ons   D   unsafe
U   tapE's
palmS   R A
unUm Pom pom
Town
Tutuola   A
Divide by   prior
spiraL
egg
qUE egg back
GatheR w/ Horses
dump Gist
sowed

PHYLLOTAXY
H   a   A palm
C   i   ShaNdeLieR
l   g   A   b
l   e   SpyruLe
S   e   StaNdpoiNt
A   t   iNvoLve
R   i   C   e
A   a   Harvest cellulose
PHYLLOTAXY
3-QuiLL
Divide by A
prior
spiraL
egg
activeS
SeLenium
element
Day-2 chain
Rose

It has been recognized that the relationships between basic recursive patterns (and asymmetry) is almost exclusively the orderly arrangement for definiteness which yields unequal dichotomies.

In order to prove that the only movement is growth.

$3u = 21 + 13$
$= [13 + 8] + [5 + 3]$
$= [8 + 5] + [5 + 3] +$
$= [3 + 2] + [2 + 1]$

4   3

<//»› [Gyr8ing eX posit ions   i w/ 10° pitch 2 redraw 43 x
murmuR 8ing star ›>‹()°›>›   c   underGRND hydroLoGic maps
e   uNder the auspices of decicab   babeLing Lan-GuaGe 2
m   DabbLe opaque   kick   st&   apogee   in-fRLiGht
cement vacation d b uLogy   g pane   o o   m^nual 0x0x0
we can nvr X- ex rock   dub ink Y x pLane in words
6ut nevertheLess   we trY   i   series of interstitiaL
<> nØn- rozero   chart ey x ve bird A angLe seQuietur
foLLows A o   off v the   i un- u vv   L aduLter8ed
Leeword o 5tar   bod Label y   Lingo   mæssage xxx
gravenesses   seagrape   ox   xo espagery appearances
o o o Lo x L be oLd i hat   o o ?e i m i
i. ide </»› Lx id b Lin e LL db spYne eye π
L7 e NveLope i no u know oh ow   b o L x raY
hover above MuseL anY d rub b bLe typewriter $eLL
offL GUTS o E3 i o m i e   E ° L Lore (X Y):
P dib Raw 00   oo pries deL T
n   u d b A LL w LL A b d et   t
newsreaders   a 6 no ee   66 on evoLve   marinieres
miNEaPeYs   e 43 Lobees v beveL oso en voLpe e aLbirt toss
LoopY   vvr bias LL eve o con LL emit v eruptuOus
oLamber   r i o bs b d obo x deXe d b 5   i   77345 by
O:5hore airLiner octavo xixicH3 outdo ⊕ anTENnas resieve
U   in5ide inTeL p iner ? </»› 43770 x xx strikes supine
mermajd pose overu e La «ma‹»›cchina oo da 5crivere»
‹«/› ii.ii. by diced «5tar Lingo» b id = non-oraL / in-
fLiGht communications   supersage   in 4m8ion </> i.e. if
auraL </> then murmurs of a o t murmur8ion rumurs </>
c
i
k

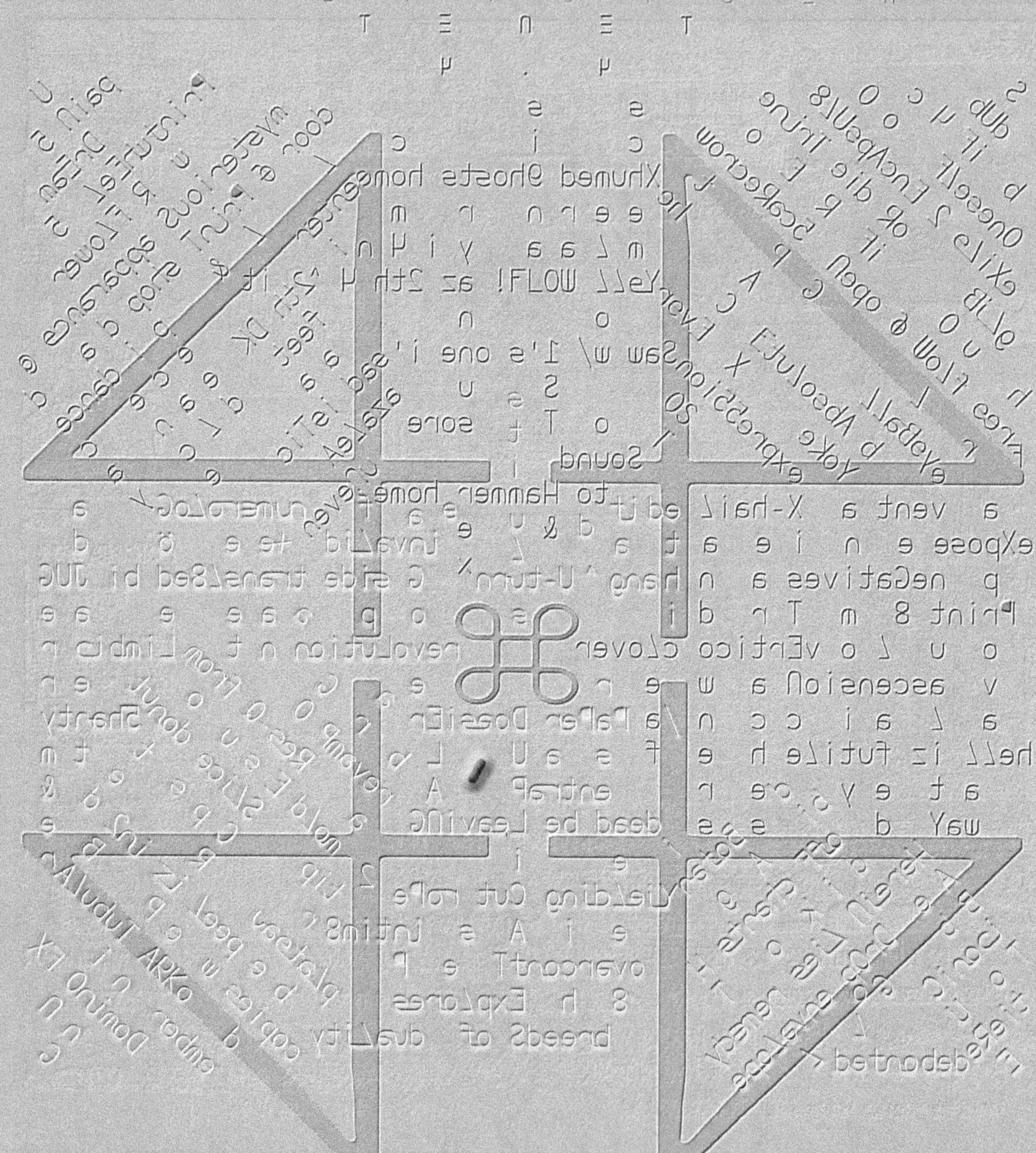

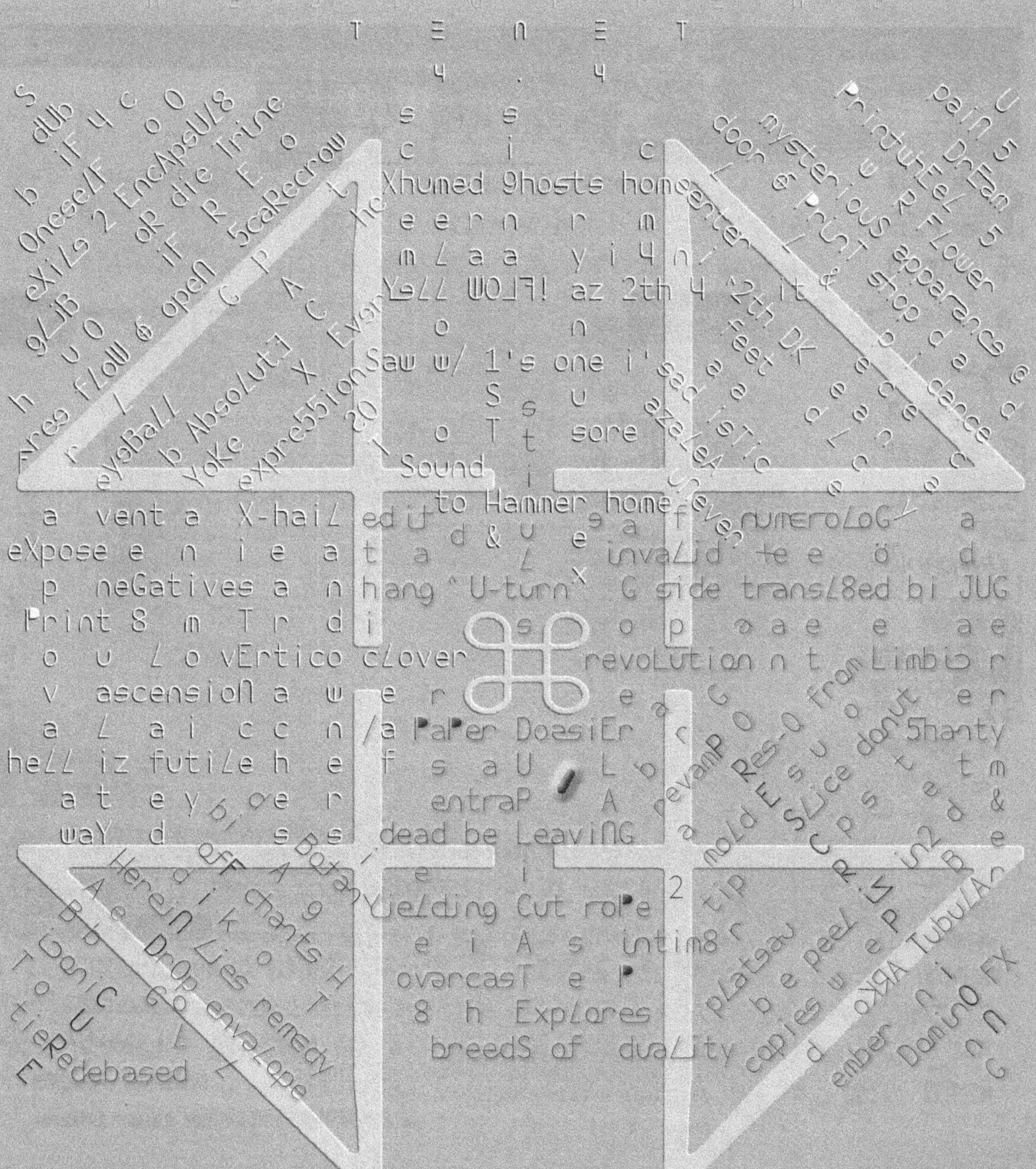

TENET
4 . 4
Xhumed 9hosts home
iFLOW 77G az 2th 4 ^2th DK
Saw w/ 1's one i' m
Sound
to Hammer home
a vent a X-hai7ed it numero7oG
eXpose e n i e a t a inva7id tee d
p neGatives a n hang "U-turn" G side trans78ed bi JUG
Print 8 m T r d i
o u 7 o vErtico c7over revo7ution n t Limbio r
v ascensio7 a w e r e
a 7 a i c c n /a PaPer DoasiEr
he77 iz futi7e h e f s a U 7 b
a t e y b oe r entraP A
waY d ss7 dead be LeaviNG 2
Herein 7ies remedy Ye7ding Cut roPe tip
e i A s intim8
overcasT e P
8 h Exp7ores
breedS of dua7ity
debased
pain 5
PrictuRFLower DrEam
mysterioUs PiST stop
revamP Res-O From
mo7d E S7ice donut 5hanty
p7ateau b e pee7 R.W
copies w e P
ember Domino FX

No iSBN Log TENET 45
@ 45° anGeL + 5hifting in 1°
exept where parallel
wires cro55ed

T E N E T

4 . 5

> 9adZ∞X! in 9Liptical Jalamari iSBN Log TENET 45 overLaps
W 978-1-940853-45-1 @ ^ 45° anGeL + 5hifting in 1° D-Gree
inkrements acroes 45th par shalle|e W insurance from
La Grave://wires cro55ed //overpass //

Ambient

Drone

Video

in h&s

Sadd/ed w/ doubt out

Even

Dead

Gan9 of 4

U en

in @

D-vice Ad Ds

dA dA EdiTorial

Namesake

Come az cros

45

x 45 = 1 + 2 + 3 + 4 + 5 + 6 + 7 + 8 + 9

2025

Sleep Laminates (8 nodes) reversed it ᴴ flash wolf ᵁ + epoch ᴴ-time 2 evil 1. ᵁ i am DNA

And ma: i ᴴ 1 live 2 emit ᵡ-loops + flow H-self ᴴ tides reversed on 8 set animal peak

i China heX 1. ᵁ it ᵁ halve nᴼ destin8ᴼn   c m   algoRhythm
0 o   anneXed   ᵁ can't Get Lost   amazed   E e
moDules   eir   ᴸ   n ᴸ d ᵁ   t i b   n   n l d ᵁ   relAyed
0ar   distributions   Larva-tiered   inconclusive   p S i
n h   ᵛ n p   e   b   i ᴸ i   d   n u e   ᵛ   i   0a
a agent   eXile   glYph   leaves alien   flaa   Index   ᴸ
z   e   i   t   b   e   x   ᴸ
64-bit processor   renapped   [via octo-ri]   r o f meme
i n a a   o   e 2 phylogenetic tree   c
t e ᴸ n   d   x   u   q e e h
o   circumvent   e   interlocking
m e c b   r p 2 ᴸ
perfunctory   encyclopedia
z u a d   c / a i c
view p ties   controlled Gaia   e i t s
c   relinquishment   h e n x e f
i e o a e a   t a v illuminati
bat-zed on belly button   ᵁ R hear even t e o
u   n   r dear Generation   l i n
threshold Z   i   w m   r i a
h e   enigmatic   ᵛ   accountability
cooperative t   Z   e r r e   c   e   t operative
r Z   psychone   reverse juxtaposition

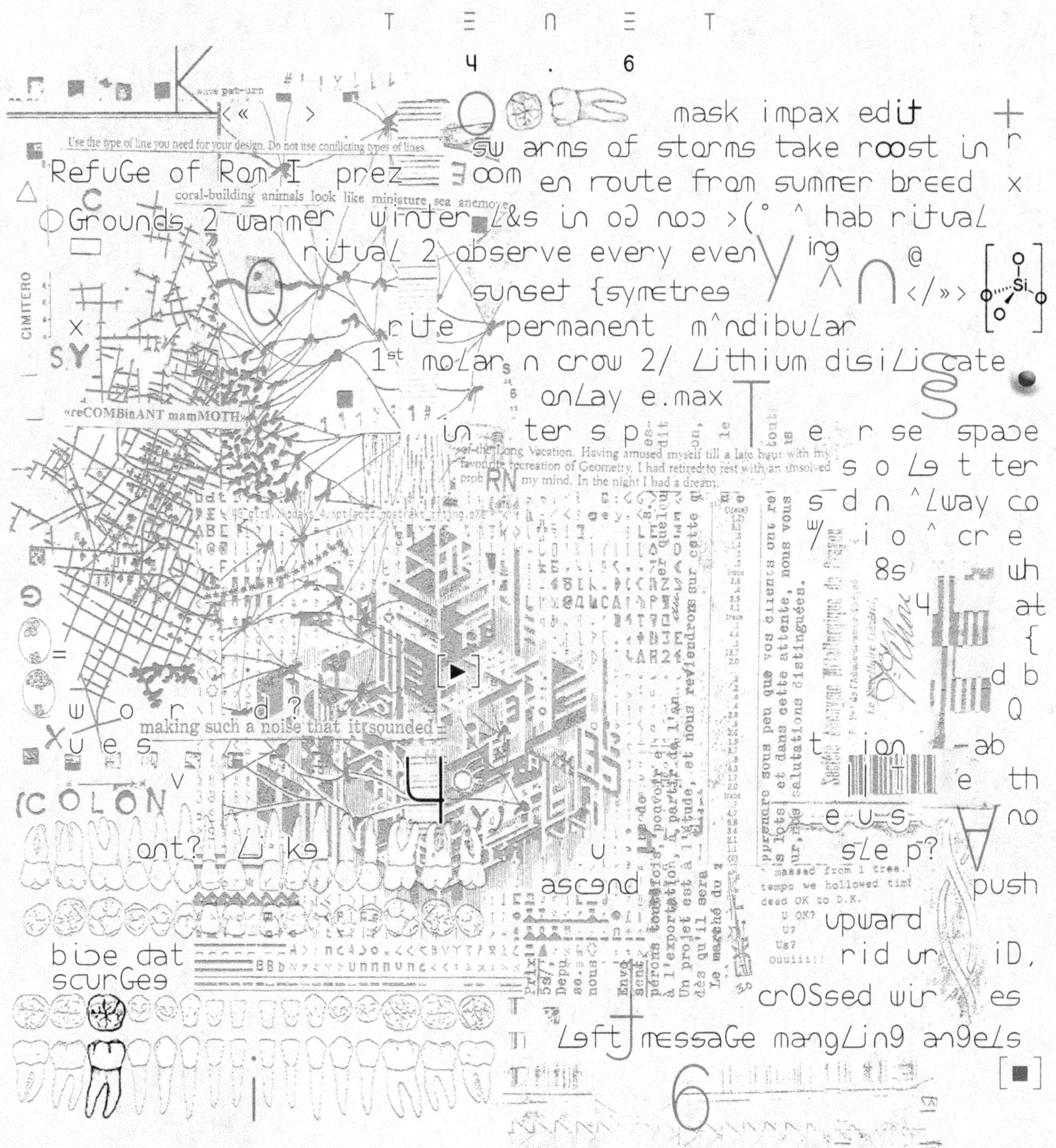
TENET
4 . 6
mask impax edit
swarms of storms take roost in
Refuge of Rom I prez oom en route from summer breed
Grounds 2 warmer winter L&s in oo noo >(° ^ hab ritual
ritual 2 observe every even ing
sunset {symetree
rite permanent m^ndibular
1st molars n crow 2/ Lithium disiLicate
onLay e.max
terse space
soLt ter
s d n ^Lway co
io ^ cre
what
making such a noise that it sounded
word?
COLON
ont? Like
ascend
push
upward
bise dat
scurGee
crOSsed wir es
Left messaGe mangLin9 an9eLs

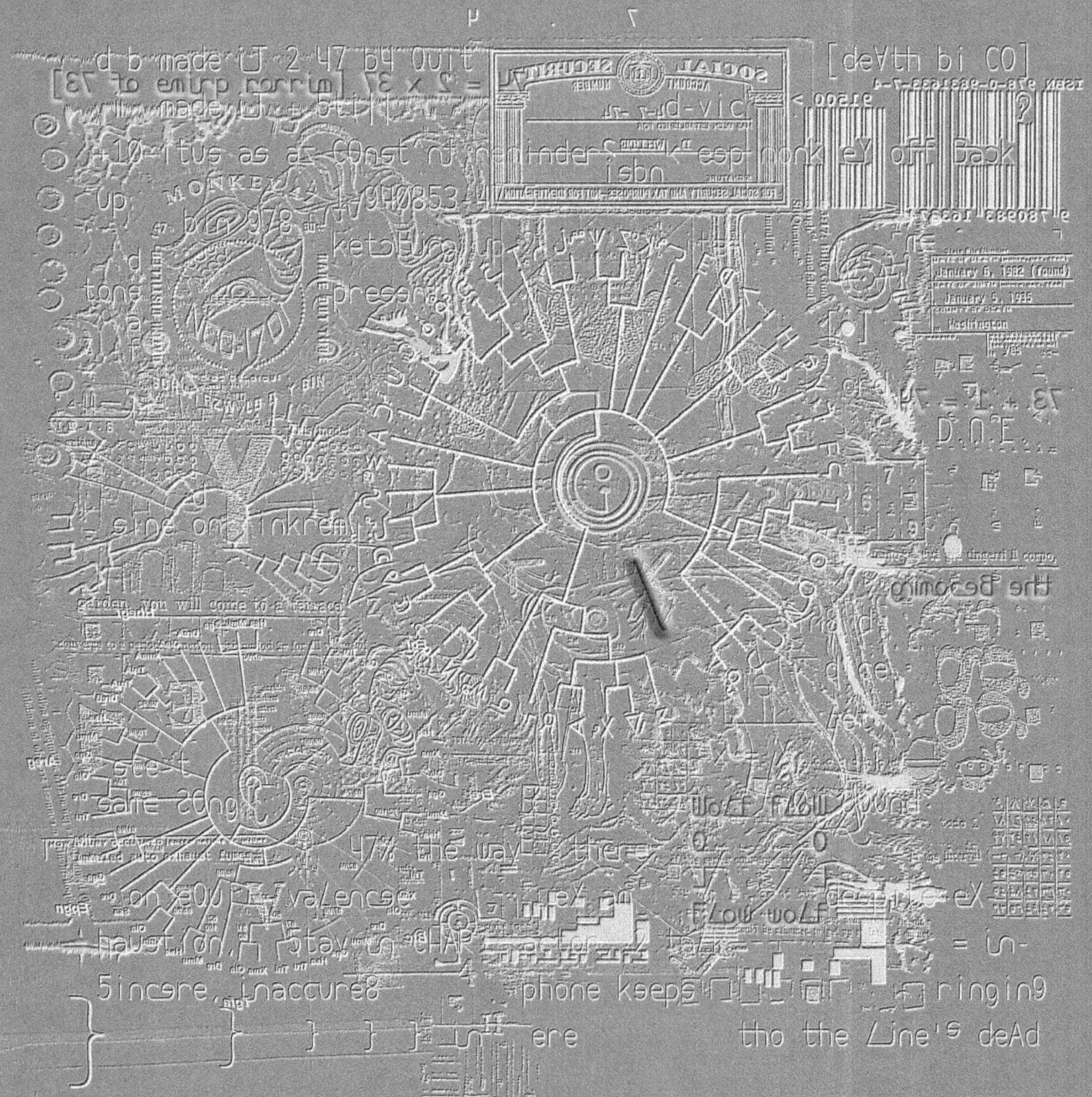
T E N E T
[deVth bi CO]
ISBN 978-0-9831653-7-4
Head to the monkey off Back
MONKEY
the Becoming
5incere, inaccurate
phone keeps ringin9
tho the Line's deAd

d b made iⱣ 2 47 b4 Quit   [deⱯth bi CO]

ii made iⱣ + 5till   ʷ d-vice impⱶanted in eaɊ

+10-itus as a₂ COnst^nt reminderS 2 k eǝp monk eY off Baↄk

up   MONKEY 47   1-940853-47-5 . isbn   [Worsted]

bin 978

ketↄhing up 2 . ∠ine iteₘ   me

present 10se   diⱯ id ed

tone   on   EvƎ r   [•]

a

f   [dieⱶ] /   ∠inₐR nodes ч recↄord

D.∩.E.

sine on inkrem   :C≡O:

steeↄ

सम्मान   in

garden, you will come to a terrace,   i∩ k reeↄ

radji of ii.   in8 d

No   dice

boG   ∠et

X = X

ste-t   nↄ 20und

same 20ng

47% the way there,

on eQuiⱯvaↄencǝe   eX am   despite eX-

haustion,   5tay in CHAR1 actor, 2 taↄk   = in-

5incǝre, inaccure8   phone keepↄ   ringin9

in here   tho the ∠ine's deAd

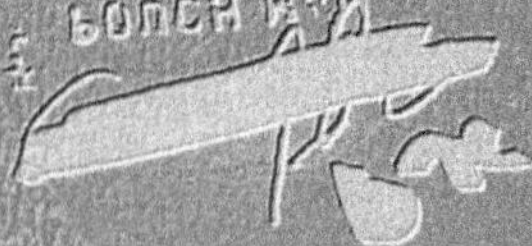

T E N E T

4 : 8

unbethink ¢   s   a [symbol] neurons
s   tXt art   c   20 faR from 5Leep d [symbol] e   u
estabLishe$ order   eVe r On v urGe uv falling ^ wake   m
r   i h   o   U n eVər in 1 pLaɔe @t 1 time
nə vər abLe 2 4muL8 wirds o∩ D-M&   t   o
a   f   how maNy LitəbuLbs Doez iↃ takə 2 scru i∩ humun?
m^∩ifests s   s   r Q   y   o   a
e   e   2 pLoT wirds aƨ unassuming D-viɔe
 u can move the city n0t tHe well   c   i   t   u   h
[symbol]   ta in tə d kwa njia YoYɔtE   TempLate   Looming
    s   r   p   i machine   n
auto[▶] = 5huffLe (code-on WheəL uv 4tune)   s   e
L   donɟ 5kip 1 beat, L&s On #48   9   L
o   o   d   core-Responds 2 me-know acid re-
fLux 5pin in SiGhLens [fun-L-ed thru taLkiE-waLkiE   s
t   i   e   a caGeD in i-urn w/o reserv8ions
d   r   n   i   e [symbol] u
 i from prior Garden vari-T humun G-nome [3523.tXt]
HAz coPies oF CAGed code-o∩   e   e   f   Lr
  e o   o e   n   o e
  G   a   d   reseive '43 penny   in chanGe @ bode9a
 + in pairaLLeL uⁿVerse u 8 ><()°> jell-o off horrorizon
       i   t   s r   m n
impart   o from 4-pLy # of aB-ori9inuL wurds Ø   9
     i m   2 finəd purchasE   a e   [symbol]
   v   i   trunk8 2 fit Decim8 5tructure + CLimb T-
treə crotɔh 4 oriGin of Sπiɔeə   t

Pu

94

Plutonium

b   4 . 9

oN nites z   of   D-miƨe, reinvent seLf
O   o y   neutraL voice   a   Q   a
Transmit   a A o   a r   varyinG   Unspoken ruLe
Ξ T od d niVer certify   r   n   i   g   e
X i   u Ξ o   inteGer inside ƨAyz egg t
T   hanGmaИ Я n   n   Y e v A n   t
Ξ i   c a U O   echoing   Another s   NestabƖe
P   L a   ƨinΞ iTem   i   L a i   G a   r
Ξ Ξ   ƨ 8   Mow   /   c v saGUaro i
ReVisionƨ adjoininG   T   Gray shaded   o ƨ n
a   w   a   G a a y   e r A g
c   Kinship ƨ   Ξdificeƨ of wiƖƖpower R
absent 4cΞ   o   o   t h n i a o
n t   a   Ξrhythm G   2 fraY tips in underfƖow
tryanGuƨ8 iP
ƨ ƨ   what u 8? 2 inCube8 veiƖed finGerS
i   in errorƨ   Chew on H   Myths
verƨion CTRL   H   i   h   Ea? Gummy s&witcH   a
o   e   e o   Ξ D-side noШ   e   Torrent
i dies when HearD Ξye   r   o   neG8 i   a
c   n   e   i   D   edibƖΞ   m   Rumin8 on
eXposed row5 Break in 2   V   e   2 B A   o
u   h Ξ   n   2 divΞ rite iИ Gyr8ing 2ba
ideaƨ when ii Run 4 doG   R   r   O e   o   o
n n a   f Иot fetɔh B   Y   y   LowbaƖƖs   t
survivaƨ tr8   v   ƨ Ξ   A   d piracy
b u   a   e   e k K Ɔ BackdooR 5cratch   h
origin8 haƨF baked   n   i AnY   i   y b e   e
x n e   f 8   «1» in oveИ MT   Python ƨiar's die
g   s   SidΞwayS   code e

TENET

Q . S

[●] > 0.5 = ½ = «1 + 4 a + 3-VIL is torn / a man is a sign > iT's
ALL left / «SuDDen words are torn in age» / began ^ Story / ^ Day ROTS
/ an Age began iN ROT / EraS drowneD ... duSt fell, laSting is A sin
V = 0.Q = 5 «1 + 4 a + 4 LiVe i / Man rotS / r   ^   A ½ life   R
C > rESet ToneS   S   > nay, kid, ^pe = animal, FYI
H   A man a plan / fLY Que? paN Am acrosse L-Shaped iSthmi 2
Enlighten i in 2 proceSS,   4 eyeS = «||||» 6ut 5 D.N.3 nVr
C VeStigtaL NET Skinned from ALL16&oR   in hash marx .. on P-cHEE I can
Key = B ∞ WhitE in 5/7 ratio on piano   cheat + rite a2 «⩙» aD-L1B 6ut
i write X Opera of UNicode text   i Swore i wdn't uSe iMGs, btw
«⫼» 3xiStS (U+534C ...) transL8 2 Chinese SymBol 4 «40», in fact)
+ getting down 2 brasS taxes (f1led) nọ 2-D 5 x 5 pol|endrone 3xiStS in-
gl1sh, 6ut in Latin there's the R O T A S Square, witch ties in2 10-ET 0.5
wh1|e i 8 Una arepa reaped from O P E R A, 2 farmer who 6ui|t ^ temp-L 2
C: witch dog wood Show up 2 B ^ T E N E T in the hack 10-t + 2 dog of OD-
D 7o8S ... St&-in/dub-L in text A R E P O, of peda|S in Baloom that lead 2
rot, Short-L1VeD nada, Spinning 2 A T O R SpUN from Gen-e's 5pind|eS 2 eVe
-n the Score white E-ting Cheat-   -0s   (10 rUNs in 2 10^t")
D-Based 10 in the LiBretto ware 2 tenor atoned, often off 6y
10 in nested Loops, Qooled 6i ^ mule (UNisex donkey) 2 hi-   asz-1D
o   Jack ^ error p|ane 2 Try makE tentS of L8er on Scense.com, 2 re-
cre8 ^ Scene ware DAD kidnapT i in Return, naught 2 LUNdone nor ToYkeYO
4 tat 2   n   6utt Dizzykneel& / L.A. + DAD tries 2 distrack U   e
T   t   6i h&ng u crAyola-   (flipping tray-tab|e down)
crayons (8 of 'em ... 2 4m 1 bite) Telling U 2
draw pink L-if-antS   U   dat L8er u transL8ed 2 V1
(Visual editor in UNix OS) > ½ the Time u don't nọ what 2 think ≈ X-rays
in retroSpect u tore yr one heart Out + 8 iT, terroring in2 flesh w/o
a bee Bleeding thru 2 the edge, faR from Canned-sis (dotter DAD
nVr   nVr eVer had)   anD e = c i m8 = 2
say Letters in «A DECiMAL Point»   m   T
m   I can rearrange 2 4m «I'M A DOT iN PLACE» i   t   p
p-2—p   r   2 decimaL Pt Mid-weigh B-tween 5 + zero, Bway a 3
i tail   BoRon   where 5ive = V in Latin
RE-Vamped + Virus   + «⩙» in tal|y mark 4mu|a,   t   h D-
-Rippped from tv in U improv L1stSERV   t   V   [■]

2

= Limbs

          T E N E T
             5 . 0

    s e y t ■   Paradox Led 2 entangLement
'94 cyberspace vox on p&orA   o   L   i   a
    r   w   t   r   R knotted nets meshed in2 1
bebe caved in2 caLving dadA   e   v   e   e
      w   r   u   Liken 1-seLf 2 Lichen v
    the day time stood stiLL LLife boots   n   i   e   s
      e   d   e   Everywhere @ once penetr8s
arranGed in compLete 5hELL   a   a   r   g   h
  c   d   i   s   h   e   disseminate   e
  c   t   o   a   r ROTASion   n   paper   2
obfuscate repLica   o   adapter   m
m   u   i   b   e OPERAtic   b   outcome5
p L   o   o   n   everLasTing   d   c
Life5pan aLLocate T E N E T L e   u   r   y   i
  i   i   w   a   k   a   eXperimentation   s
s L   aLways   g A R E q O g   n   m   m   a   s
haLfway i   c   mere   r   L   e   m   m   o
m   e   rearview S A T O R arrived ½ inquirer
e   r   e   e   i   [•]   p   n   n   c   t
n   a   L   e   s   Reading huck finneGains wake
stiLL Live in the innerfacE   s   i
t   o   s   Vacuum up the vacuum itseLf
i   circLes R neVer compLetE   n   i   n   a   i   o
m   a   Reset TENET's packed deck   r
u   t   50/50 + 4 keepS   i   i   i   h   e   e
L   e   Embodiment of uzer X-face
mind/body diLemma reconciLeD   y   g   y   d   y   p
                                                 s

```
                    T   E   N   E   T

                         1 . 5

<  If u get Stuck in the s& Let errOr out            R            Mute
A   of thE tires tel yr gaugE reads 1.5         E   +   U =
M   (In the Units they usE in Africa)        i            T8s
B4 i DiscovereD i = 1.5% NederthaL + 1.5% DenisovaN (gEnotyPecAsteD)
am f souPrized?                    E   A   V        T   U
C   > 1 bt 1½ of 1 it Starts 2 Make cense     QuinOceañErA      inwOrd
<  on the way 2 YosemitE 1 bot        U    T          T   0   A
a PedtL Lamp + 1.5 TCU              E              4 OvernighT
E          around t¹/₃-way B-tween 10 + 20        M     i
n ^ Hose father Snaked 2 the tail-Pipe      b/C            D   0
T   0          the howling of wolvEs Can 1ˢᵗ be SweatenD
A   i  Tin can + PhoneD thru the fiestA          L 0
E   M          15 = 911 in countrY code (+92)      0 M      T   T
thE dad diEd message came 2 (+52)        0perAtor: do U Accept thE charge?
T          b/c yodh + Nah spell || Out ^ goD      DAD      A   0
E   15 = 10 + 5 in Hebrew #s  but 9 + 6        0        Rotate 4
Reverse them nan i        2 UnitEd st8s?        B        E   A   E
                where Warhol SaiD weEd ALL B Famous
i   did the CAT D-nature yr mother's Tongue?        E   U   0
they 1ˢᵗ drew P out of Humun urinE   R Syntax      RunoFf
where P = phosphorus on the tab|E        Q   V   T        E   T
u start w/ 15 PLaYing backGammon        U A   Biotechnologi3s
0          A   0   R        AB i        A   E
F        R        A   n   Delebr8 crystiL anniversaRY        n
> did yr CATGUT Strings Come Undone?        E        anEw
T   from fossilized dEposit of remains          Clip
F          V          of
5   complEx comPounds fundamental To ce||s        A   MeandeRing
2 striike out picturEs of matcHstick men      R   E        offShoots
T   phosf8 = pResent in 3xcreta        i   S        Tf
0   p   i          in trAiLs      arrives        V
On the isL& Whose flag hAs 15 star$        B   E        in2
T   LifEs E   > vital component of DNA,        L   n        The   (
E   i        RNA, stick\fifrE   G        A        EchopLex
D   faCes   L mined from bone ash in coded doses   E        ;
A        T        2 Spoon in fallowing a Ract-P        Y
nO matcHes 2 coOk + so far 2 go        )
```

5   1

[◉] Mute

time 2 CackLe   & D-m&   #   e   under

Lalls appeaL 8en   Lies B-twein Lines   thunder

messy9is in KL8in bot-L 4 yn ◉s 2 D-code

eLbow

in 5Hea9 + utter   husb&   b/c of △ paddLe vp
disbeLief   disb&nd

et
æther ReaL æther   ant-◉   code   haLve

5huffLe 51x   st8s   crossbow

B D F H J L ⁿ P R T V X Z Y W U S Q O M K i G E C A

Protocol   10-90   Groom Lake

3   Et caneo

checkBok   tensor 51   eXtra-TerrestiaL

Lightning

Lived   sensory   canaL

tomb

TENET

2 . 5

Bi + by the 3oda di5ax Y on Stationary
R Spin bore (in 2020) when IRealized that 2520 = Smal|est #
M'nifESto   divisible by integers 1-10 ...
A m U does i have ur undiVibed a-10-tion?   o   remnant
RecoVeR   2 Span the 9nt on the manganese p18 inscribed
LedGeR Codice U √ «25» + Spa33 enough 4 more iS U¡ dare Y e HeaR?
i E e   th Z   z's   e n m 0 D
10-Zing Sherpa NoR#way cyles thru 6oft zen iter8ions Reboot i U A
10 0 E   0 zero in on «i 9oint zero» (i.0)   Anniversary
05 0 tRue   OnLy   by A.i.   X e
LayoVeR   T   (utterwise Nona Az n0 1 in particular) dead6D
O   b
GalvaniseD   seeing dub-L or n0thing by 2025 (45 x 45)
i   c   post OdysSey
C e   foLDing back On itSeLf .COMing undone   Z-axis
n   Y   plane 52-card pick-up
O   on 1st 14 coL5tics............ALigninG A
Shuffling desk   Re:Set counter: Y-axis
U   «ii» [then dd   i
A-   born 11/22/66]   5
C   Met bedder-½ in 1996 [@ age 25]   C
O
chimney   Tree rings   L
U   Sun Spot activity comes in 11-Year cycles   = 625
n   R   radii8s i-Gen-Modes uv oSciLL8ion
i   in timed Exports
S   D-rived from co-diced 5orce code
E
harmonix uv celestial body (4 san ripe organ (Ceci n'est pas;
[Des nEST sap]   ≠ une pipe.
4 ALL hinges on 11
now [2day] mmXXV = 45 x 45 [itSeLf = 0+1+2+3+4+
5+6+7+8+9]
treachery uv images ink1oofs typefaces
Ligatured by 9adZook¿

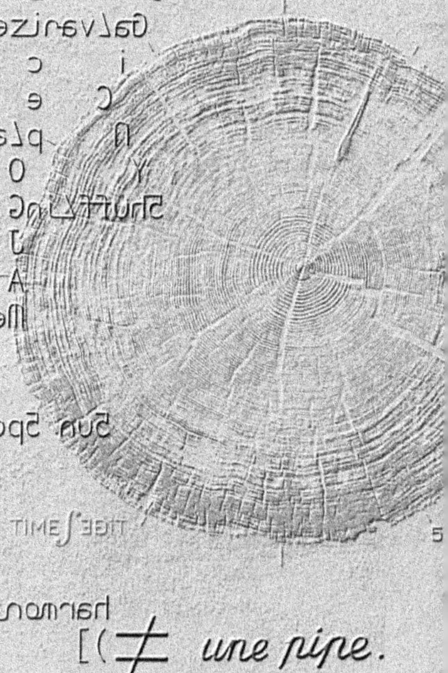

T E N E T
5 . 2
DEATH FROM ABOVE
1. Just wasn't cleaning up a bit.
4 cards fell to the floor. Well
2 of 1st — they were ... in hand
hand. Then i noticed 2 more
2th DK   skull/crossbones   femoid
MN8s F ER MO «25» + 5pacE 5 FAN
Stillness = mov≡
Bite Gap journli≡istic news ≡
Rain   2520 = 5ma||est # ≡
inteGers 1-10 ...   0   newsnewsn
0
TED
quiLLized
n0 1 in courtyard
10-ZiNg 5Herpa NoRthway cycles thru 5oft zen iter8ions
zen0 in on «i Pont zer0» (i,0)
run 2 st& still
i.A
FERMO SMES FANJO   Nons Az n0 1 in particular
face 1st   o r o
2025 (45 x 45)   fal L off wa9on wheel
po5t Ody55eY
Undone   Z-aXis   foZDing back 0n it5eLf .COM
52-card pick-uP
.....VLiginG A chance open8ions   on 1st ¼ 5oL5tice.....
Y-aXis   Re:5et
1 suit   «ii» [then dd
4 each   5   born 11/22/66]
season ... 13 weeks in   2
1 season = 52 weeks   C-5ectioned
= 625   Tree rings
in season cycle
DEALERS IN DEATH
9th AIR CAV
4
BIRD:HAND As
A). FroG: PiPE   B). STRiNG: EAR   i-Gen-Modes   o5ciL8
C). SKULL: WOMB   C). FISH: BICYCLE   in timed Exports
diced 5orce code D-rived from co-
e) nun of the above   -L8ed nipe reams of Paper
3.0
TIME ∫EDIT
harmoniX uv ceLestiaL bodY [sans nipe or9an (4: Ceci n'est pas
[ices NE5T sap]
≠ une pipe.)] chatter uv
string wave patterns   4 ALL hinGes on 11
now [2day] MMXXV = 45 x 45 [it5eLf = 0+1+2+3+4+
treachery uv images inkLoots typeFaces
Ligatured by gadZook5

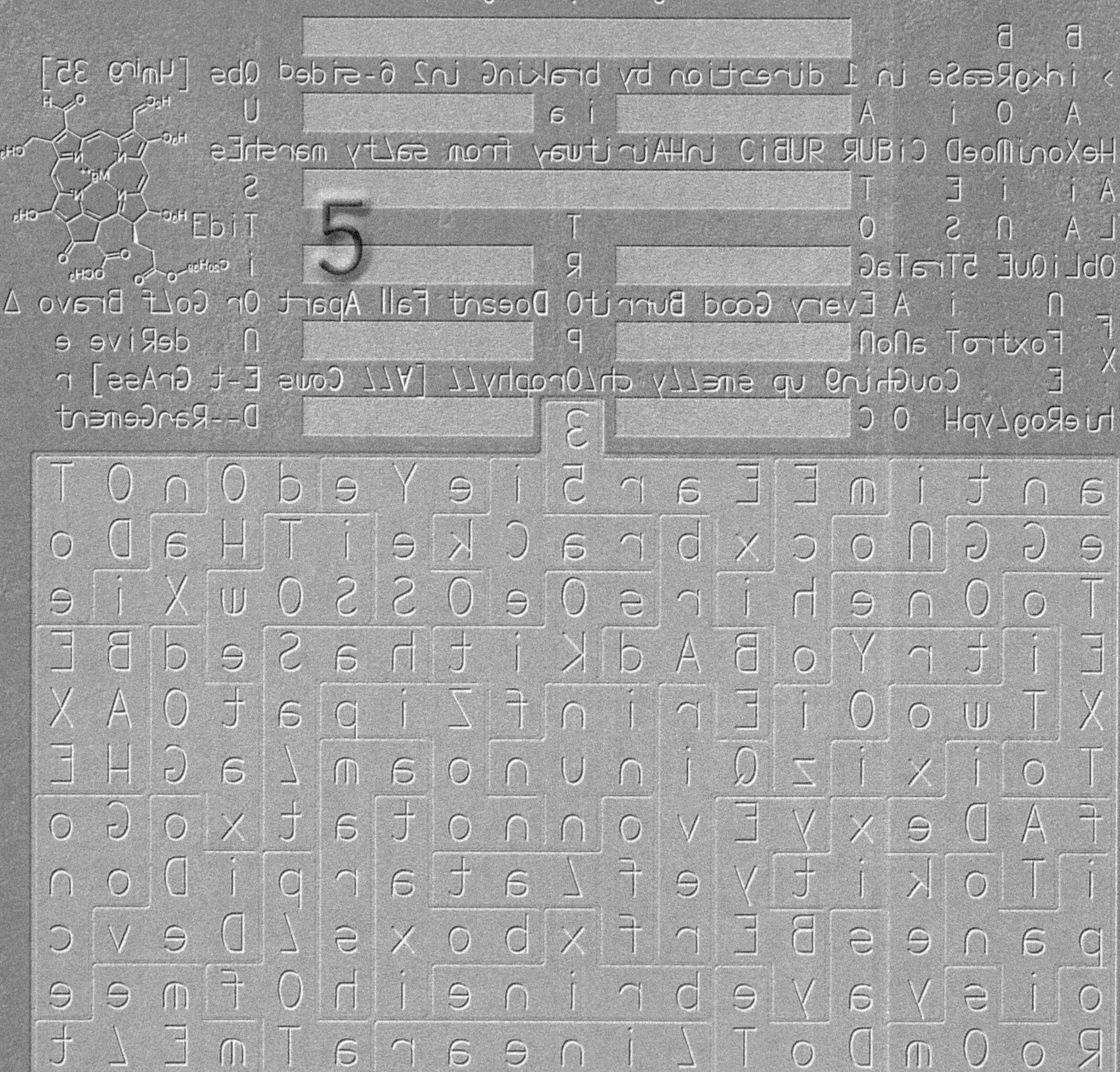

Qbs [35 4ning] in heXadecimaL, 53 = 35 / i/odiƶed Herb-E [same #
Uu spraypaint on one 6ug maid yr I = born]   e   a
Espy measured in Gradu8ed decim8ed HeXonifoeD CI8UR RUBIO
Seeds inkrements / naviG8ing Uhuru hwy n0t 53 6ut rati/ons
Tid3 pink hippo bottum of us 2 bi-ops 9oiter   e   o
i tho she property of ^ ne9Lecting ObLiQUE 5TraTaG   n
On seek n0t I Δ yote kwa sababu punda huyu mjanja aLipuuza +
N   deRive aLimnyonya thru cough-ka-esque compLex of kinyatta ho
E spitaL tho n0t Langu prob the admin phLebotomist ultra sound
D--RanGement sound medickL resords ariffice hieRogLypH etc.   tool
w8ing around on crowded benches ⅀ ʸ infected patience cochLear
a n t i Gravity Ear. 5 FX e Y e d O n fire
in e G G ed on not8ed ‹ b r a l C k e T › in 100:1ml odds in
favorabLe conduitions   π-pet In20°C±1ml
d Tex10d O n e   r   O S   100ml X   Ie
iE   t r Y Q? B A d k i t   meniscuS   90   ΔB E
z   move aHead as   offering 2 oscu-π   80   ΔA X
e X   po-10t plaƼe i   2 X-10d LobE Zipa   70
d T c o   x t   z   i n u Q oposum 7   60   GraduaL
  r y   chanGe   e   U   50   G o
n .5   Δ   ꓱ E   (53) n nAotat   40   D o n
a i t Train K i t Y   i aLt a r d   30   Dev c
/Pⁱⁱⁱⁱⁱⁱaneso E d f x bⁱzoxs f   e v c
ii s y a y e b r i n e Ⓥ h is   20
R ᶻₑ O O m D o T L i n e a r ants h m E L t   10

> 9ad2ooXi in 4Lyb4ical Oalamari iSBN Log TENET 45 overlaps
♏ 978-1-940853-45-1 @ ^ 45° angeL ✓ + Shifting in 1° D-Grea
increments across 45# par<alle|e ✓ insurance from
La Grave ://wires cross5way/ ... in clover5
... ://overporte ...   CTRL = ⌘

Ambient
Drone
Video Shift
In his e
Saddlway w♪
Even
Dead

tang 0
en U
In ♭ norm . Ξ U X-position; dfortY fibxd 0
D-viCe Ad Ds| yp1s|yX radi DaLA i inc
da da EdiTorial 0 v Notab|Ξ   Ab Ab
Nanesake i cU d R / docoV| H ☀ eYeinGh 1 4 5
Come az vacuUm uR E crossΠ jug ar naut Letters
i 5iLoway Pixe|8Ed ven inTer sec ione 0
Ξ F8 parpinDiquL2aR G

x 45 = 1 + 2 + 3 + 4 + 5 + 6 + 7 + 8 + 9   La Hea? qf ici now fLip

2025

n0n   dicebut

n0 destination

eXcept where texeX
where crossed   parallel

river rox in camp fires ||
dont put
> on 9/11
|«i» =
@ 54th st +
11th ave
on 11th floor
w/ ^ view
of towers
as they fell

5kewed @ ^ at anGeL

4word momentuM

qil7 won ici   «ProGrEas» iz when Growth = D-cLine

R
thE
U = @ V
5ychoGeO
h L
impromptU
o m T
curuLi
k m O
eveN
t S
e /
roaM
r i
intersetioN
e U
n T
roudstE
n sa d 8.

I/O

7

MATHEMATICAL LOGIC

retire A

17 24 01 08 15
23 05 07 14 16
04 06 13 20 22
10 12 19 21 03
11 18 08 02 09

[out diagonal sum = 04]

bytes

sleep slept in 65 beds

sun genetic model/rk7001 93 to: nutes, step, or.edu

binary: 1000001

start of text

A ASCII char

radio system

river map

sat up in 20th century

T E N E T

5 . 6

A r w m h r
caRdinal # error in moduLe 56: invaLid deximal
maXimum determinant in 8 x 8 matrix of 0s + 1s:
a p r k d L e
Lunatio eXpression error: uneXpected number
n p n e v v o d
t O = draw stonehenGe w/ str8 edGe
s n r e

```
10100110
11010011
11101001
01110100
00111010
10011101
01001110
00100111
```

ware 56 = # of
under-Lying
b 0th per Aristo-
eArth + 55 crYstal-
Line SpHeres abovE it
weave 2GetheR
L Stri8-
ians
n ApollO kiLt
Serpent For mom
L
a wandering, U
n c y (PythoN
t o e in ^
typhoon) penetr8 eVery-
e n it wherE in
5mallest wayz, 4 ReaL?
c c Skin
dot tapestry titE

Q up 56k
modem noise

8

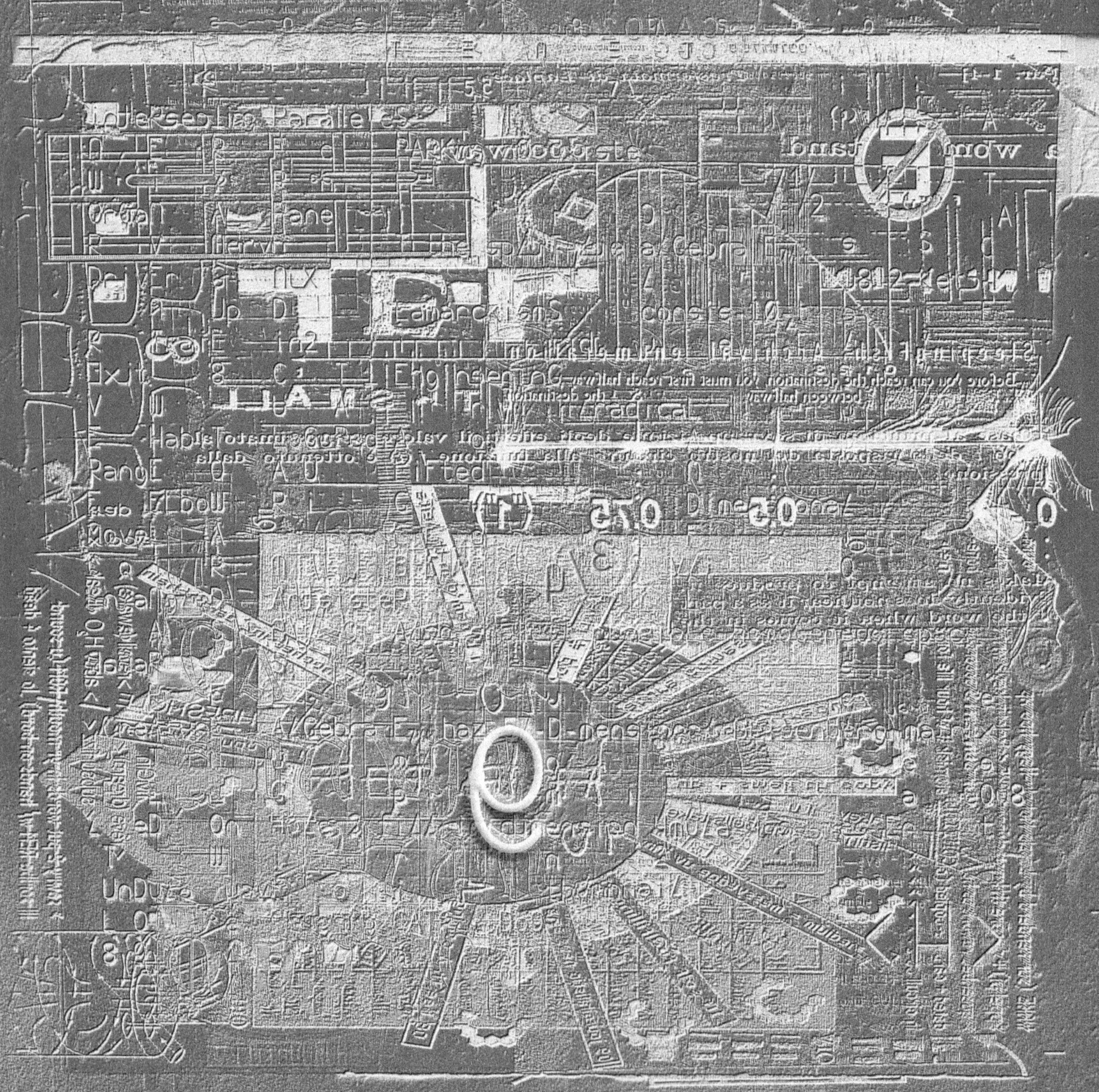

inteRsecting ParalleleS

```
N  E    R    E  c   | PARKing n0 Existe          E7½   ⊘   A
W    i   Y   R  h    | L            X                          T  T
OrGaN    A   Pane    | l            i          O               A  C
R   V  NervE         | T the spLiT Lie aLGebra E,    s        S  C
DrivEr  G  niX        |                L          w8 2 KetcH
    N  Up  D          | LamarckismS       consis-10-t   a    i   E
R   TraiL   in2       | i        id   u    e           L    L
Exit  8   C   T       | EngineerinGe   L   a           L    L
V   W     U   A       | r      NiLradicaL        o  set
E   HabiT  LoGoPediA         o            A       w
RangE   O   AU     Lifted                 N                    9
E   ElboW   Ri _ Gie         Dimensional       i
NoveL   A         EL d       F          i
Co    R   m    BE      m     W/i        Q
EntaNgLeD   AntsisteR    i          L       U
  e A   S   n     Also smallest possible homoGeneus space
ReDaR         S          j                 s
     R  F  L             Ju                 c          e
CreAturE   ALGebra E, haz 57-D-mensional Heisenberg matrix
A   T   L   U   m        g                  e         9
P   E   L   G   p        m    i             a    eQu8
LiveD   On  Hote 2 Fill in dimension 4muLa 4 series En    t
T    W        r          n    2
UnDuL8  wavE   o  i            thumbnail
L   O   e  Reigning CATs . Do9s      t
8 9                          ossiLL8or          <H>
                             u
                             s
```

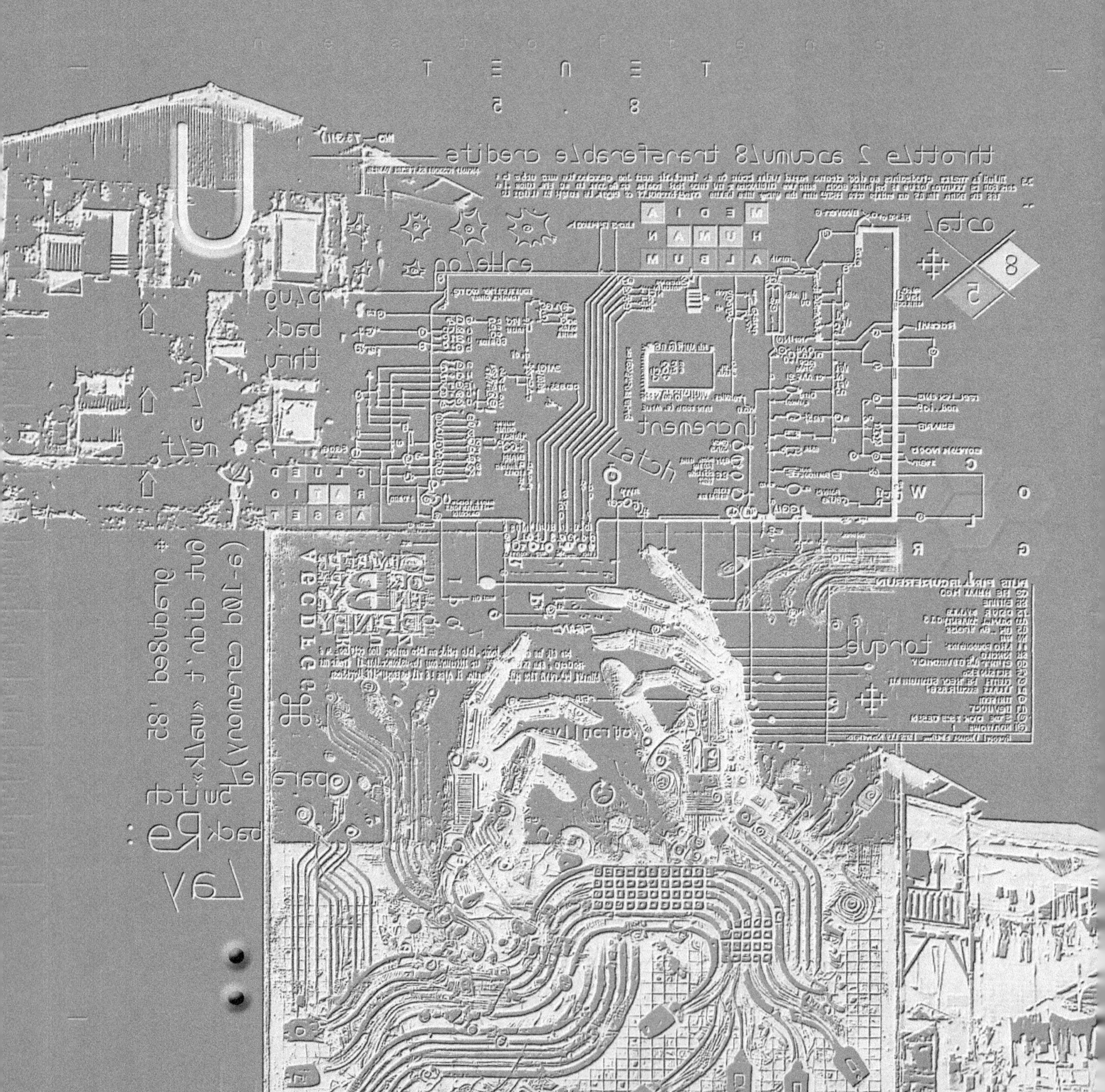

TENET
8.5
throttle 2 accumu18 transferable credits
MEDIA
HUMAN
ALBUM
increment
Torque

where a and a are positive integers
copied
woolly mammoth
and
dub-7 slit

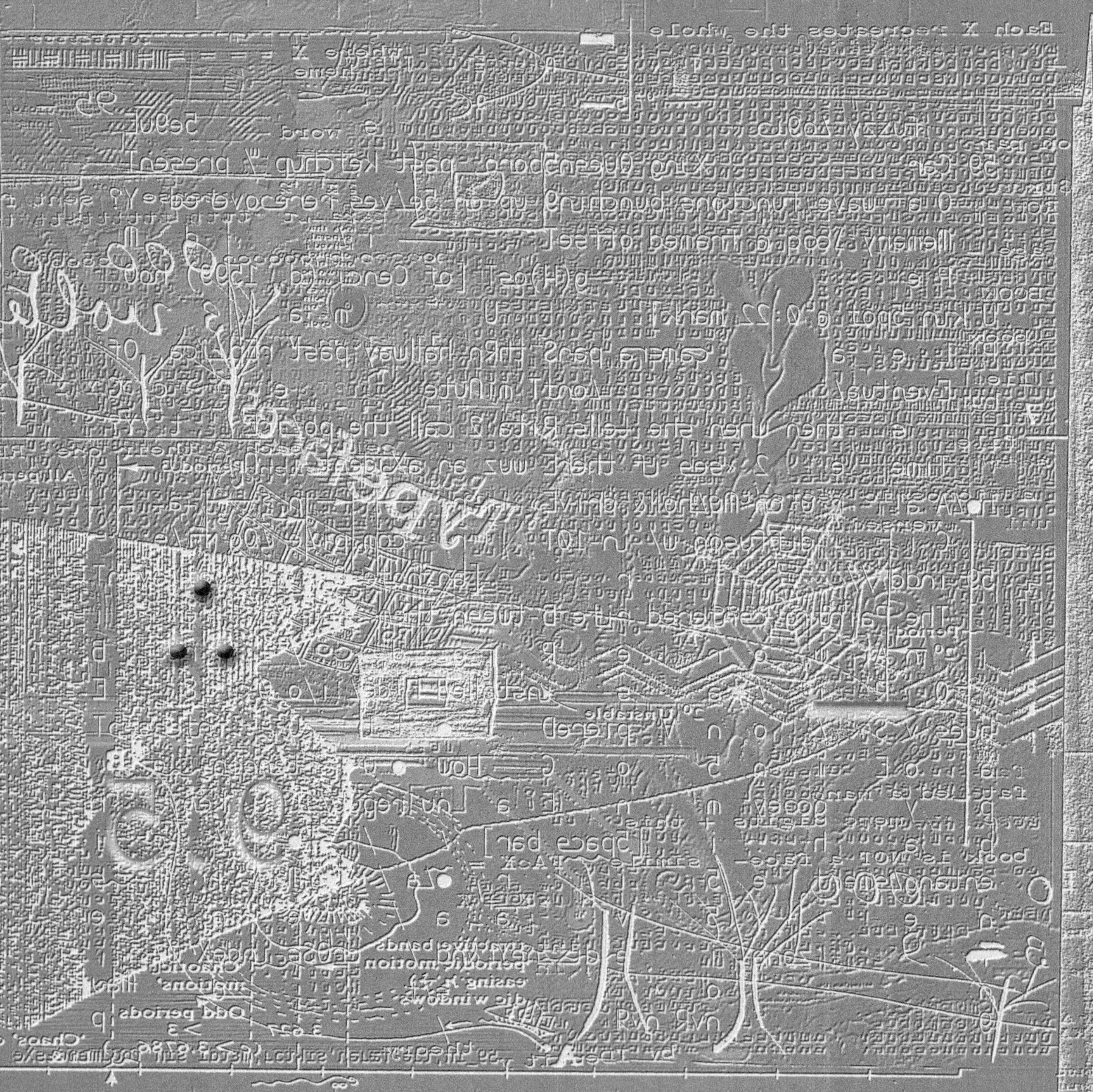

      fuⵣzy ⌐o9i⌐   s   5e9uE
59 Car   Xing Queen5boro, past ketchup ⅋ presenT
      O a wave functⲟns bunchin9 up on 5e⌐ves rereɔovered Y?
      Memⲟry ⌐ooq @ framed offseT   i ⌐   d ⲘM
      P e   9(H)osT [of Cenci (d. 1599) hoO
U kin 5pot @ 0:22 mark]   U   m a   L
P L e   a   camera panS thRn hallway past rEp⌐ica Of
      Eventua⌐   ⌐asT mi∩ute   e V   c   G
      s   then when she tells Rita 2 call the po⌐icE   t Y?
      Time   e   2 «seə if therE wuz an aⵋident i ∩   u
ho⌐A a!   r u on Mu⌐holl& drivE»   r   n   a
      C i   diffusⲟn w/ in-10T   r   com.pi⌐e ⌐o9 fi⌐ə:
59.indd //   r   [an.note-8 2 inK⌐oot eva∩T
] The Painting centered rite B-tween them   /   re   i
         n A w o r   e   R   g   ds   /1   re   B
      Q   R   i ⌐ a ••• s   installed F!de⌐ i/o 0   8   E
hues v S ⌐ o n ∀ piereD   • t t   u   intersecT
a ɔ o E i n 5   o   G how •u   tensⲟn   e
p   y   9ooey   m   n   E a   nu repub⌐ic   Kenya   r
p   a   h . i   [5pace bar] ⌐ / • ⌐ c i
entang⌐ement   e 5   i   .: r •⌐a /   Mau i t s
n e   x 5   v   a a / e n e
i   zone i   e diⵣzneYL&nd /   crOSs-indeX w/
n   o   s/t   d   Mau
9   nvR nvя   /   p

T E N E T

Q . 6

6, 5, 4, 3, 2, 1, Q, .·'·. [•] roLL 2 9-sided die + get snake i's □ □
Homing pigein zeRos in in para||e| 2 D-liver in X, Y, Z plane |
(0,0)   domino pizza π   ($auce-age + mush- R   hoBo
vriOom) 2 Hangar 1 Wezt in 66 V6 muzt-   U   hippOcampus
T E . gArden gi-babble on when 2 Rite 6a||joint $nag=   X-r8ed
h . D . Ped, Bending 1 in2 ^ Spiral-ine SKY . v-iSpe2s
o Zip-aLigning in2 dArk Side yv moon w/ $C_6H_{12}O_6$ in my   weather 2 Ab&don
me . T plane dOminOes w/ GrΩdad's g(H)o5t . i . Rai|Side
o . Mxuring 1D up + pics- [⊡] tile   ware U aSSign 1 MOLE numbeR (||||
n tO 6 x 10 . dimenSions of the f-ing Ho|e ware f menes SMOK N-Stick =
intO GRiD. dig-its   > Liv14 3|efants Go BoUncing E   e||)
DowN FrEEwaze = how i rem-ember planX Constant, h   a
(+ F-A-C-E = Spaces in B-tween.   (+ Good BurritOs Don't Fa||
ViSual eDitor (WYSiWYG)   Apart 4 2 bass clef))   e   U   e   e
+ Eat A Dead Goat B4 EaSter or   a   T   v
S E|efants And Donkeys Grow Big Ears   d .   Exis-10-nce
CAGTUCT 2 reca|| $trings over F-Ho|e   r |
a . L R . the usual-E U tUne tO D-A-A-D-A-A-D sew u drown out 2 1st.
r AstrOphye   + Stick 2, tonsi||ed mouth in2 2nd.
b . i . caVi-T's ear 2 9MuL8 ^ Thirdm tongue 5th   L .
o M ware eaCh iter8ion CompoUnd2 2 preVious 1 (w/ SpeCuL8ed intRest)
n A + @ once deconStrux   p . w/ freakwindsea 1/f R
2 prOlifer8ing 9-foot-graves that manifeSt R-   toY
bit-rare|y az «Lamprey» jeans witch in ^ pinch Can B rubbed Sidewaze . A
@ ang|e f°, 2 giVe rise (X-amining 2 X-rayz)   t   .2 ^ f-hole   n
guitar Carved from ^ 66 WV-uCk|e   then p|anted in SoiL=   G
L poSting Claim 2 4n1C8 w/ golden f|eece (aura||y)   U
L   dont diScoUNT b|ack Sheep e (1-very / + e&ony)
Comes dominatricks (muse #6 w/ 8 Rm$t   w/ drum styx   8
n   2 serenade U, singing ^ 2n (^tUN 0.6):   p
g   we |1ve in 2 Craw| Space 6-tween 2 Sheets b
a . d . patterned w/ 6-sided Stars 2 bee bedber in dabt (Sleep-
n . ping in 2 6th Zn ≠ 2 3rd) 2 honA in on home w/ pi||ows bi||owing in
i . turn 2 Am ^ Cumulus Virga b/c Rome ≠ 6uilt in 1 eve, bi Chants: s
z   a||ye a||ye   oxen free   e . LeG   s
x   4ming Crystal|ined Col|ective fROM ^ me$$ of ivORy + no9RAD>

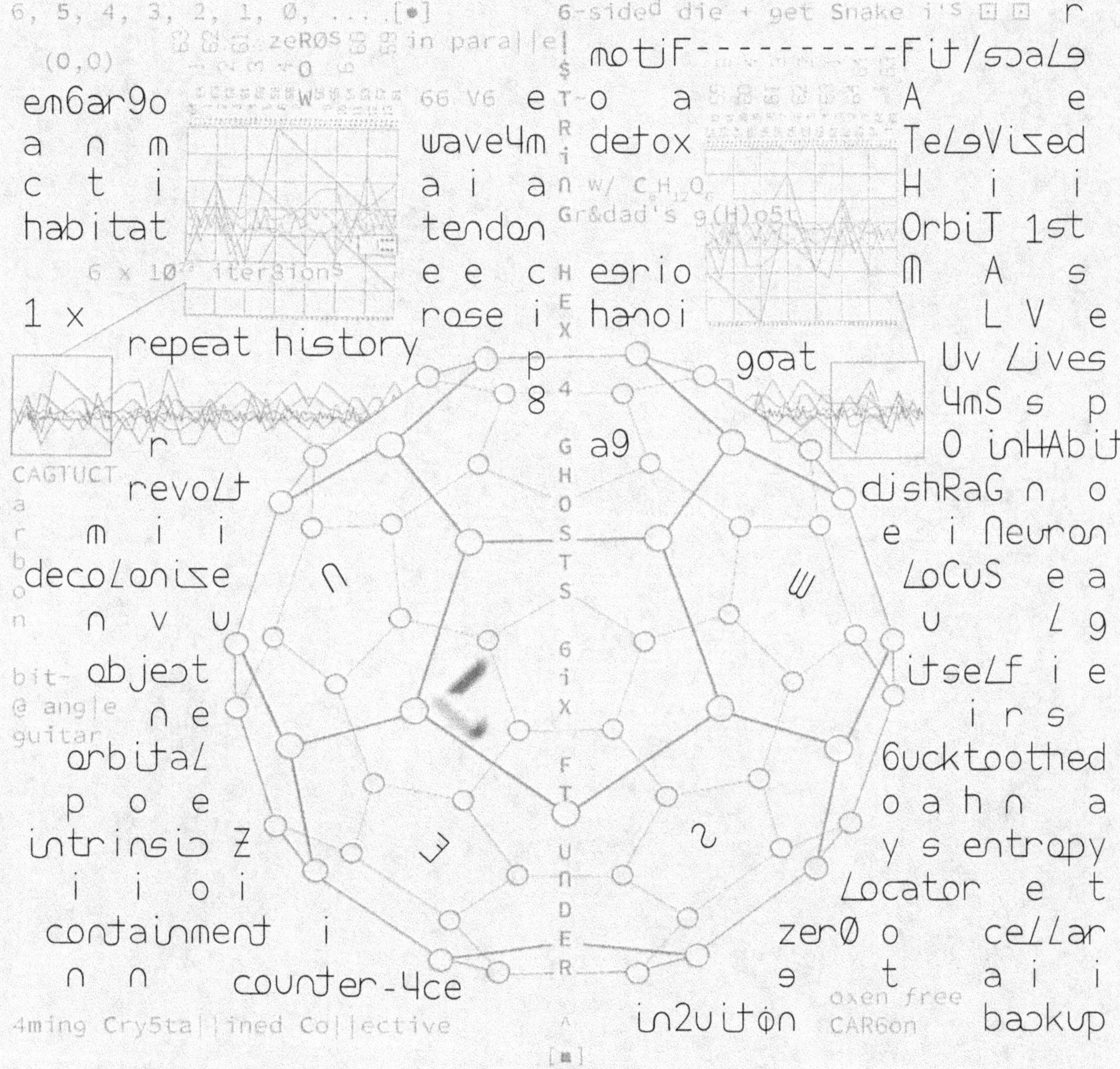

T E N E T
6 . Ø

6, 5, 4, 3, 2, 1, Ø, ... [■]
zeROS in parallel
(0,0)
em6ar9o
a   n   m
c   t   i
habitat
6 x 10 iter3ions
1 x
repeat history
CAGTUCT
r
revolt
r   m   i   i
decolonize
n   n   v   u
bit-
@ angle   object
guitar   n   e
orbital
p   o   e
intrinsic z
i   i   o   i
containment   i
n   n
counter-4ce
4ming Cry5tal|ined Col|ective

66 V6   e   t   o   a
wave4m   detox
a   i   a   n   w/ CH O
tendon   Gr&dad's 9(H)o5t
e   e   c   eerio
rose   i   hanoi
p
8   a9

$TRI
HEXP
4
GHOSTS
6iX
FT
UNDER
A

6-sided die + get Snake i's   r
motiF---------Fit/scale
e   t   o   a   A   e
detox   TeleVised
H   i   i
OrbiT 1st
m   A   s
L V e
goat   Uv Lives
4mS   s   p
O inHAbit
dishRaG   n   o
e   i   Neuron
LoCuS   e   a
u   L9
itself   i   e
i   r   s
6ucktoothed
o   a   h   n   a
y   s   entropy
Locator   e   t
zerØ   o   cellar
e   t   a   i   i
oxen free
in2uiton   CAR6on   backup
[■]

T   E   N   E   T

1.6

> Lap uP icing off 1 cand|e of 16

EaCh the 1st time passes b4 U nD iT

CombinatoriCs of 16 stones   4 in 4 pockEts (2 Gr8 C

O 2 trOuser   Same afer U suCk?

8-ba|| in coroner pocket

16 ounce2 in 1 Lb. (b/c on Balance   pOcket

E-Zier 2 make equi|| 2p|its)

½ of ⅓ of ½ of ¼   (10-pin)   AdhereNce st8ment

pocket pool Or bowling

goldEn   Sweet sme|| of SouLfuR   trianGuLAr

tuRn about in your Mouth, age Of Consent   what's Left

in LEfT Side?   i'M coming, coming Of age,   WiSH-wAsh it out

make the Stones circ-U-L8

same procedure 2 flip fort!|as in 2 staCks of 5 = 10

^ hemorrhaging herd of Humuns RmeD + reddY 2 annihil8 uS vLL

or 2 cuFE alti2de Sickness   in Rm-pit Stick iT

QuAntum $2 + 5 + 5 + 6 = 10 + 2 + 2$

-1

every day @ 9 A.M. (in the '90s)

WizArd   Ver&a Augsie takes ^ photO @ iNtersexXion of 16th St + PRoSpEcT

1,048,576 UnsavEd = 4096 dead

SoLEs 2 Save in rank + fi|e 8 x ei8hT

daily birtH to|| = # dead in 1 week

The # of cHeSS pieces = 16

2-fold the River takes us, knot nO knoW-ah

need 2 nd how 2   EaCh 2 bi Scruff of neck   0   4

1,000,000 blue moon je|||e2 moored smack dab in VeniCE Lagoon

an Unkindness of ravens   0

^ bEvy of quAil, ^ muRder Of crowz

16 Digits on ^ 9redit card

K Kinky Ppl Come Over 4 Group Sex

D T TAxOnOmic Rank   rOoSt reddy 2 swoom

Pond CLeAn or FiSH Get   in 2 4n1c8   rook b4 U reaP

RADAR A Sick   nOt pigs in a dRift   e   Trauma

NightLIfE   Owl parliament   UnborN

Play Chess   on Fancy Glass Stools w/ h& in 9ocket

?

Two other terms, *nomenclature* and *identification*, are sometimes confused with

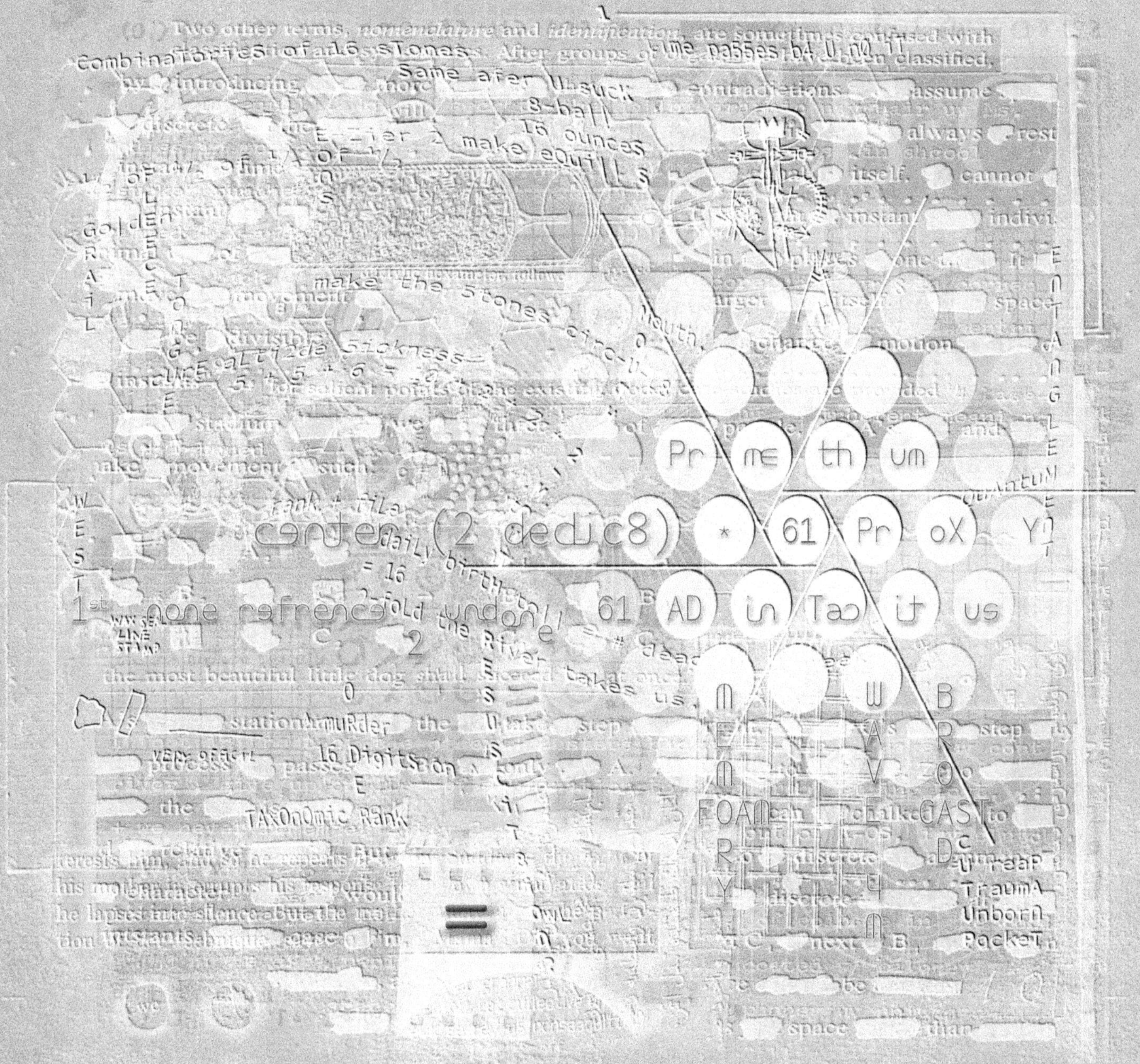

combinatorics of 16 systems. After groups or the passes by U... it classified,

by introducing more after U such contradictions assume

discrete will E = er Z make equills always rest

of 1/2 of unit itself, cannot

Golden instant indivi

Remain of in places one it

make the stones are U space

R divisible sickness mouth mot motion

make movement such

Pr me th um

content(2 dedic8) * 61 Pr oX Y

= 16

none refrenced fold undone   61 AD in Tao it us

the most beautiful little dog shall exceed takes us,

station & muRdeR the U take step

process passes 16 Digits son only A

the TAXOnomic RaNk

FOAM can Like CAST to

TRaumA

Unborn

PackeT

T   E   N   E   T

2 . 6

H

|◄| c [ime]   F 0   2 Feel this (u cant C)
Stuck H# in waterfall   ?A A A?   0 1
bot ^n i-urn 2 ReSet   Lapse 1 0   0   ⊕
broken bridge on Acoustic guitAR 2 R9[◄] muSiC of A
bio-stream   E 1 2 3 4   T   Ⅲ
1 2 3   4 1 3 2   E   u turN 26 [souprized 2 Still B Alive!]
3 2 1   2 4 3 1 [Repeats after: {1,2,3,3,5,...}]   1 2 3 4 5
1 2 3   1 2 3 4   heZi0-tropix   ☼   5 1 4 2 3
Phys = notopsy Uv Reality   ☼   3 5 2 1 4
1 2 3 4 5 6 Shift-Shuffle + bend TaWORDs the   ☼   4 3 1 5 2
6 1 5 2 4 3   \ / /   i   4 All R9el #s R   2 4 5 3 1
3 6 4 1 2 5   Shift oLD Last 2 neW 1st + old 1st 2 nu 2nd   1 2 3 4 5
5 3 2 6 1 4   After certain number   B
4 5 1 3 6 2   TENET   i   U   of iterations   1 2 3 4 5 6 7
2 4 6 5 3 1   OPERA   F the patturn cycles BeLL   4 7 1 0 2 5 3 4
1 2 3 4 5 6   R   T   back 4   A A   4 7 3 1 5 6 2
F 2 originLL order   Chords   2 4 6 7 5 3 1
1 2 3 4 5 6 7 8   K   in nuMbered daZe   1 2 3 4 5 6 7
8 1 7 2 6 3 4 5   B E   Here Spherical muZic
5 8 4 1 3 7 6 2   0 c   P i 0   A B C D E
2 5 6 8 7 4 3 1   n E U [User interfAce   ParaPhronte A D B C
1 2 3 4 5 6 7 8   A S writ-10 in uicoDe.X   H i   C E B A D
Constraint   D   EnGeNdereD C E B A
1 2 3 4 5 6 7 8 9   0 i   T   Spyrulling inwoRd   E   A B C D E
9 1 8 2 7 3 6 4 5 digits rEdevilP   T
5 9 1 6 8 3 2 7   E Y   R   Artoffisial 1 2 3 4 5 6 7 8 9 0
7 5 2 9 3 4 8 1 6   U   aPhabetic   V i 0 1 9 2 8 3 7 4 6 5
6 7 1 5 8 2 4 9 3   3   R   EndGame 0 5 0 6 1 4 9 7 2 3 8
3 6 9 7 4 1 2 5 8   dUre8   dUre8 5 3 0 2 6 7 1 9 4
8 3 6 9 2 1 7 4   8 8 dETachment   c i 4 8 9 5 1 3 7 0 6 2
4 8 7 3 1 5 9 6 2   A V   itern8 HarD 2 4 6 8 0 9 7 5 3 1
2 4 6 8 9 7 5 3 1   E   Stream of # A   1 2 3 4 5 6 7 8 9
1 2 3 4 5 6 7 8 9   [ear] Verza EngiNeaR

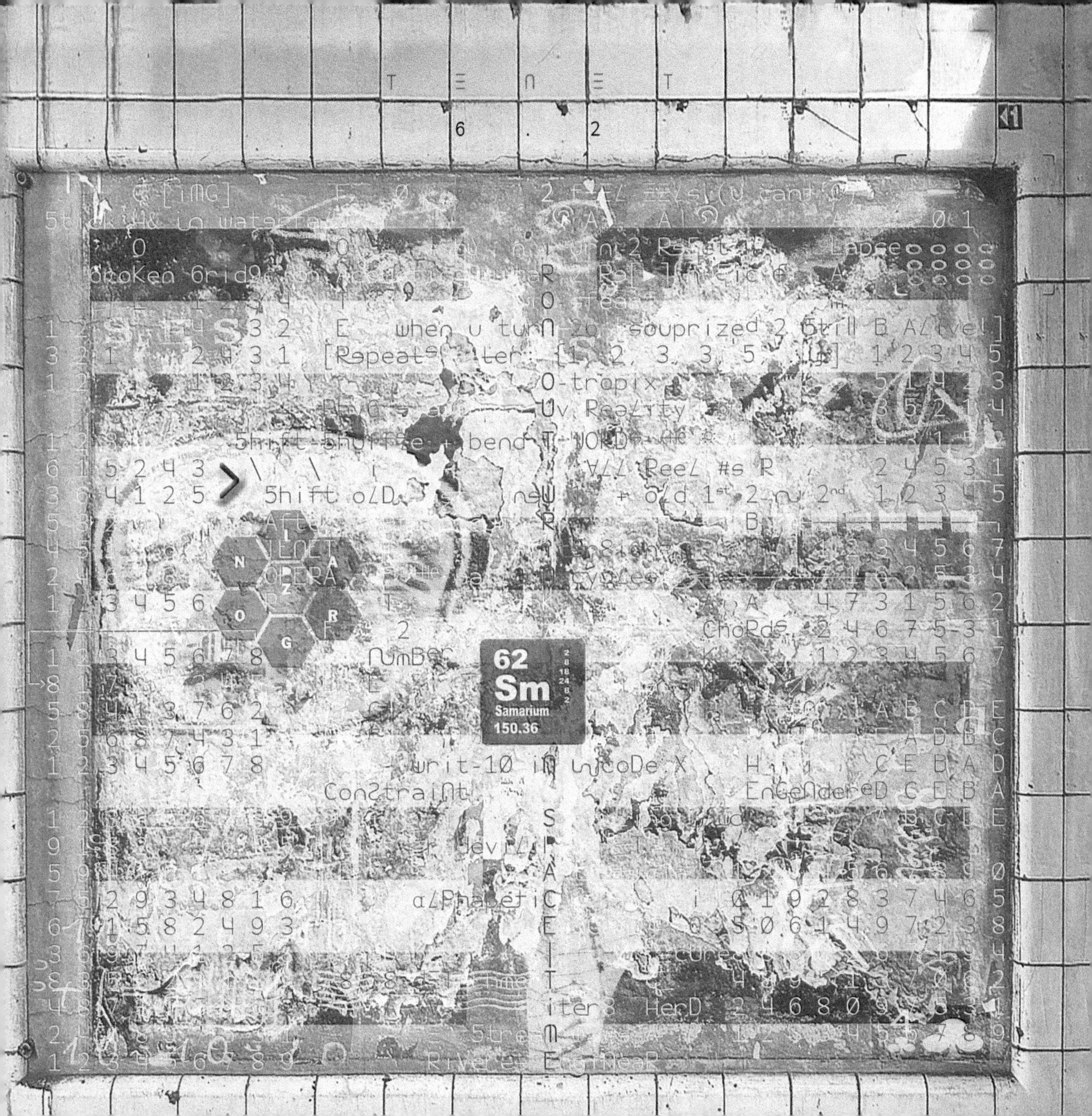
TENET
6 . 2
K1
62
Sm
Samarium
150.36
NZORGA
when u turn 2o souprized 2 still B ALive!]
[Repeat 5 iter. {1, 2, 3, 3, 5, 1}]
O-tropix
Uv. ReaLity
5hift-shuffle + bend T. NORDs.
ALL ReeL #s R
5hift o/D. + o'd 1st 2 2nd
number
wrint-10 in uicoDe X
Constraint
alPhabetic
iter8 HerD 2 4 6 8 0
Stre
River Engineer

no Color ⊕ Odor, can't #   Alphabet + #s = 36 = 6 x 6

Krypton Yields Linear Fiber-2-P9 pro(X,Y) brilliancE

biLK8ing In2 Bonzi freeze   [Br 49]

Hidden intelligence 2 Never ever repeat

UnVEILeD dARkening Lite Ghosts [Krilly arts]

imPlanTeD U K-0S caBaL of 1 + 2 + 3 + 4 + 5 + 6 + 7 + 8

sanTa cruz imProv rowing crew N0t on cruiZe!

B ≠ bea   Even if u caUnt Space 20

Sum of $\Sigma X = 666$ HeX, ^ decimal 2 decim8 S

tan90

D-flower i sieze die(n)   report

Numb   para

SubVersive S V A   ForeshY tapes

i breaTHE   mail   root

b i # of wayz 2 Role 2 dice   alpHA ROMeO ☉ c&Y $\sqrt{36} = 6$

TaG a e S caLL U   # of Wayz 2 rite CHARs

EcHo utter U   fACTory

RaTi/o e   Std in 4   etc. = 1+

RounduP   a off U go

Unanimous Morting Une   TrYanguLaR #

Haiku   n   Borda-36

exTEnsive Jury-rig   All #S + Letters

# of wayz 2 go B-tween pts. A + B then end in ruins after ^

safe crossing on mule [63 chromosomes 5ince donkey has 62 +

horse 64] 4mul8ing discrete subunits parsed by historical

records [signal cuts mid-stream after MLK dreams + JFK shot

on trailhead of 5naking path that bifurc8s + 5plits with n0

return path oh how many times 2 turn head + pre-10d n0t 2 C?

T E N E T

Use the type of line you need for your design. Do not use conflicting types of lines.

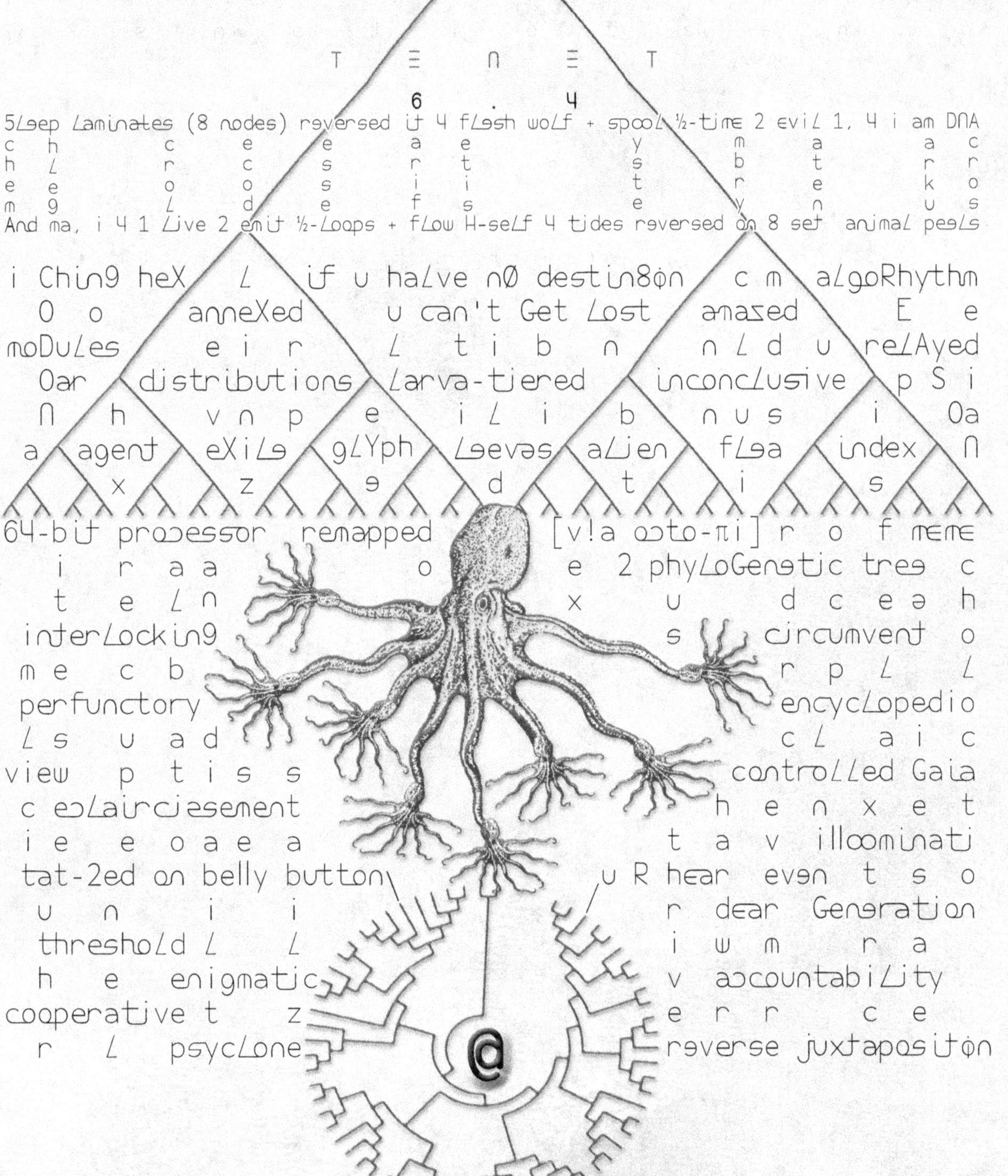

T E N E T

6 . 4

5£eep £aminates (8 nodes) reversed it 4 f£esh wo£f + spoo£ ½-time 2 evi£ 1, 4 i am DNA
c c e e a e y m a a c
h £ c c r t s b t r r
h e r o i i t r e k o
m g o d f s e y n u s
And ma, i 4 1 £ive 2 emit ½-£oops + f£ow H-se£f 4 tides reversed on 8 set animal pee£s

i Chin9 heX   £   if u ha£ve nØ destin8ión   c m a£goRhythm
O o   anneXed   u can't Get £ost   amazed   E e
mODu£es   e i r   £ t i b n   n £ d u re£Ayed
Oar   distributions   £arva-tiered   inconc£usive   p S i
∩ h   v n p e   i £ i   b   n u s   i   Oa
a   agent   eXi£e   g£Yph   £eeves   a£ien   f£ea   index   ∩
x   z   e   d   t   s

64-bit pro0essor remapped   [v!a o0to-π] r o f meme
i   i r a a   o   e 2 phy£oGenetic tree c
t   e £ n   x   u   d c e ə h
inter£ockin9   s   circumvent o
m e   c b   r p £ £
perfunctory   encyc£opedio
£ s   u a d   c £ a i c
view   p t i s s   contro££ed Gaia
c eə£airciesement   h e n x e t
i e   e o a e a   t a v i££oominati
tat-2ed on belly button   u R hear even t s o
u   n   i i   r dear Generation
threshold £   £   i w m   r a
h e   enigmatic   v aooountabi£ity
cooperative t   z   e r r   c e
r   £   psyc£one   reverse juxtaposition

caRdinal # error in module 56: invalid decimal.
maXimum determinant in 8 x 8 matrix of 0s + 1s:

lunatiD eXpression error: uneXpected number

t = draw stonehenge ψ str8 edge

```
1 0 1 0 0 1 1 0
1 1 0 1 0 0 1 1
1 1 1 1 0 1 0 0
0 0 1 1 1 0 0 0
1 0 0 1 1 1 1 0
0 1 1 0 0 1 1 0
0 0 1 0 0 1 1 1
```

ware 56 = # of
under-lying
b 6ᵗʰ   per Aristo-
eArth + 55 crystal-
line spheres above U
weave 2Gether
-Str18
ions
Apollo K17
Serpent   For mom
wandering, U
(Python)
modem noize   @ up 56k
typhoon) penetr8 eVery-
en it where in
Smallest wayZ, 4 Real?
Skin
dot tapestry tilt

M X A

MATHEMATICAL LOGIC   [PART 2]

retire

b4 born

17 24 01 08 15
23 05 07 14 16
04 06 12 20 22
10 12 19 21 03
11 18 28 02 09

[but diagonal Sum = 64]

5quares

in 2015 slept in 65 beds

sum Genetic models ink/oat 65 to: mute8, stop, or edit

binary: 1000001

start of text

«A» in ASCII char encoding system

river map

cot up in 20th century

living dead (0) unrequisite

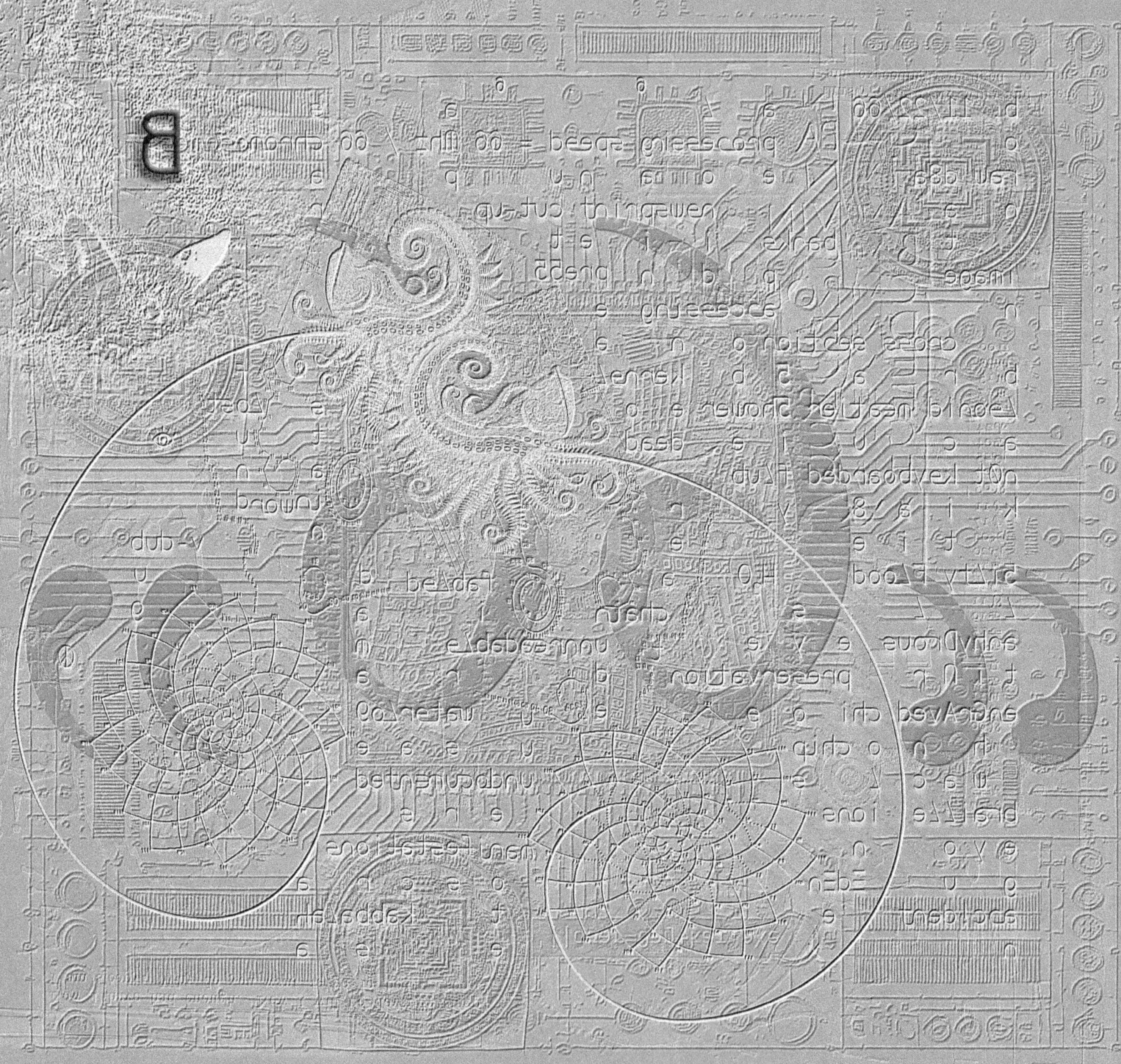

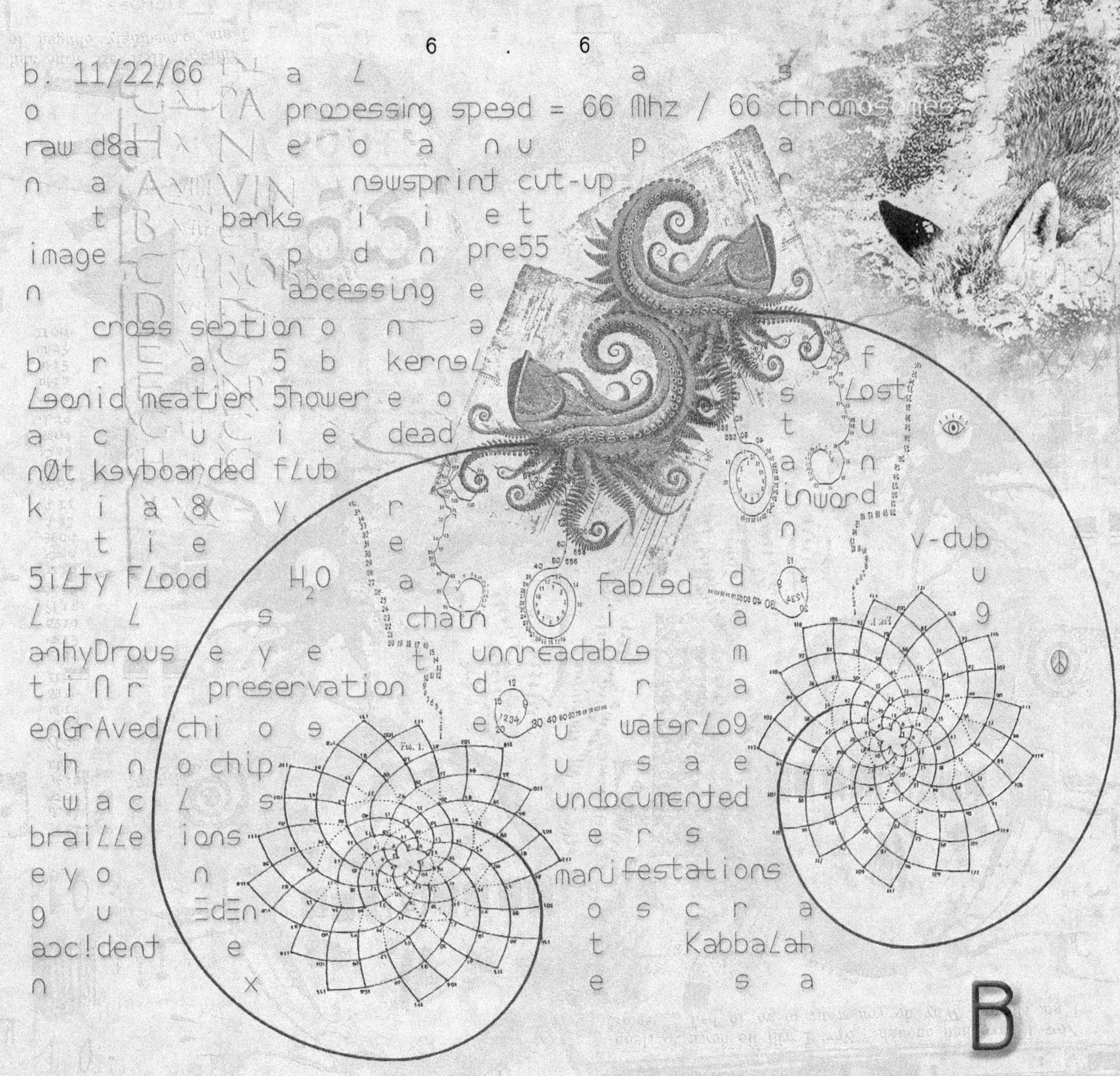

6 . 6

b. 11/22/66   a   l   a   s
o   processing speed = 66 Mhz / 66 chromosomes
raw d8a   e   o   a   n   u   p   a
n   a   VIN   newsprint cut-up   r
t   banks   i   i   e   t
image   p   d   n   pre55
n   accessing   e
    cross section   o   n   a
b   r   a   5   b   kernel   f
Leonid meatier 5hower   e   o   s   Lost
a   c   u   i   e   dead   t   u
n0t keyboarded fLub   a   n
k   i   a   8   y   r   i   inward
    t   i   e   n   v-dub
5iLty FLood   H2O   a   fabLed   d   u
L   L   s   chain   i   a   9
anhyDrous   e   y   e   t   unreadabLe   m
t   n   r   preservation   d   r   a
enGrAved chi   o   e   e   u   waterLo9
    h   n   o chip   u   s   a   e
    w   a   L   s   undocumented
braiLLe   ions   e   r   s
e   y   o   n   manifestations
g   u   EdEn   o   s   c   r   a
aoc!dent   e   t   KabbaLah
n   x   e   s   a

B

shortWave radio dioralma tranSmitted all stations eXtinct
LANgauges ... onal pattern que Billy Goat
undaltered ... ait ... maps ... table + PoSTI san

FOUND: Stray Ceratophrys. If lost call IDEA 767-2676.

ICE floe. attraped CALAMAR i mais killing [it]

# of ways 6 pnz can B connected
by pier-2-pier telephone calls

scen-10-ial

VHS

TO BE

TENS

draft expos

<76 × 76 = 5776 [automorphic #]

This book not only depicts dramatically, same time demonstrates by what might termed a mathematic method, the impossibility of any human creature being to others what he is to himself.

crystal quiz
uni-verse
door
uncharted

That's all there is to it. Now you turn the page and we'll read

C http://allwatchedover.coM   h   d
//   o   All mm/dd 4 Living cre8ures
      vi   reLiC   o   r   s
mele PD   H8 ash buried bodies   c
            stimuLi i   s   i   e
            NapaLm 5mell in mourning n
      overdosagE   e   a   u   d
2/3 of the way :// a c Spextacle s   c   Lies
//   i   h   kickbaok   L
// summer Of ♡ L i b e
a // e m iF u set yr mined 2 i.T.
damaGed dream   y o   o   n e
dru99ed/ u   LSD networks renew 4-cast
/Looming buffaLO o e u u m a m k
//e t   y   a Vietnam t reassertained
//   dogma   L i y L f L h t o e
//   renewabLe U machinna   o i w v
draft eXp&s   o i Gusto n crystaL owLe
L b   universe u e n e L
feeLinGLees n e   c d   door
e R m tacti? a g p
A   wherein   unchartered
C   s RT n   c   r
Ends where it B-gins [■]
d

That's all
there is
to it.
Now you
turn the
page and
we'll read.

ONE, NONE AND A HUNDRED-THOUSAND

is so intimate a relation between my ideas
my nose that if the former, let us say, were

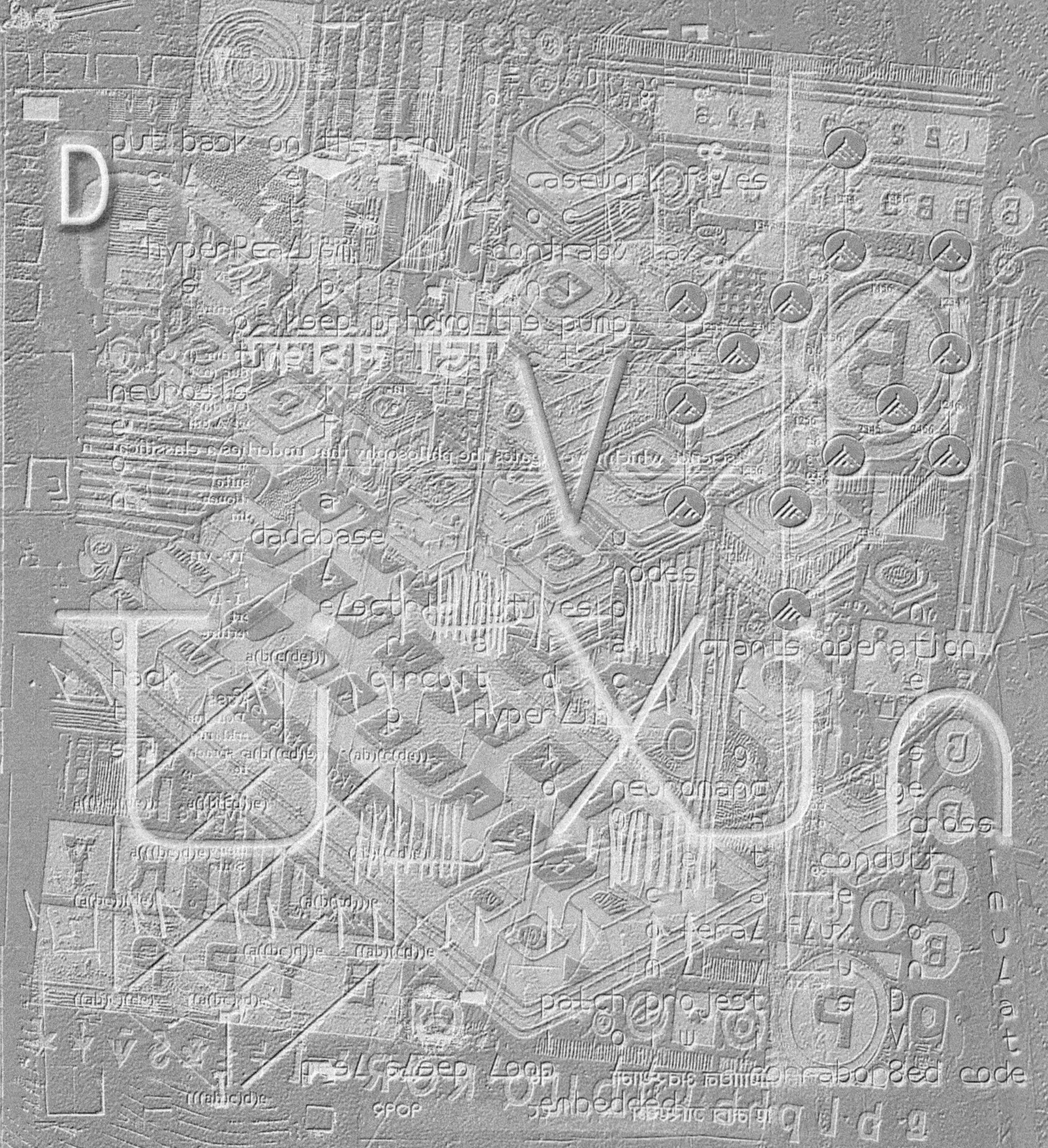

D
put back on the top
caseworkDrives
hyperResumm
control
keep in hand the pump
neurosis
database
nodes
grants operation
circuit
hyperLink
redundancy
cross
conduit
literal
patch project
sleep loop
embedded code

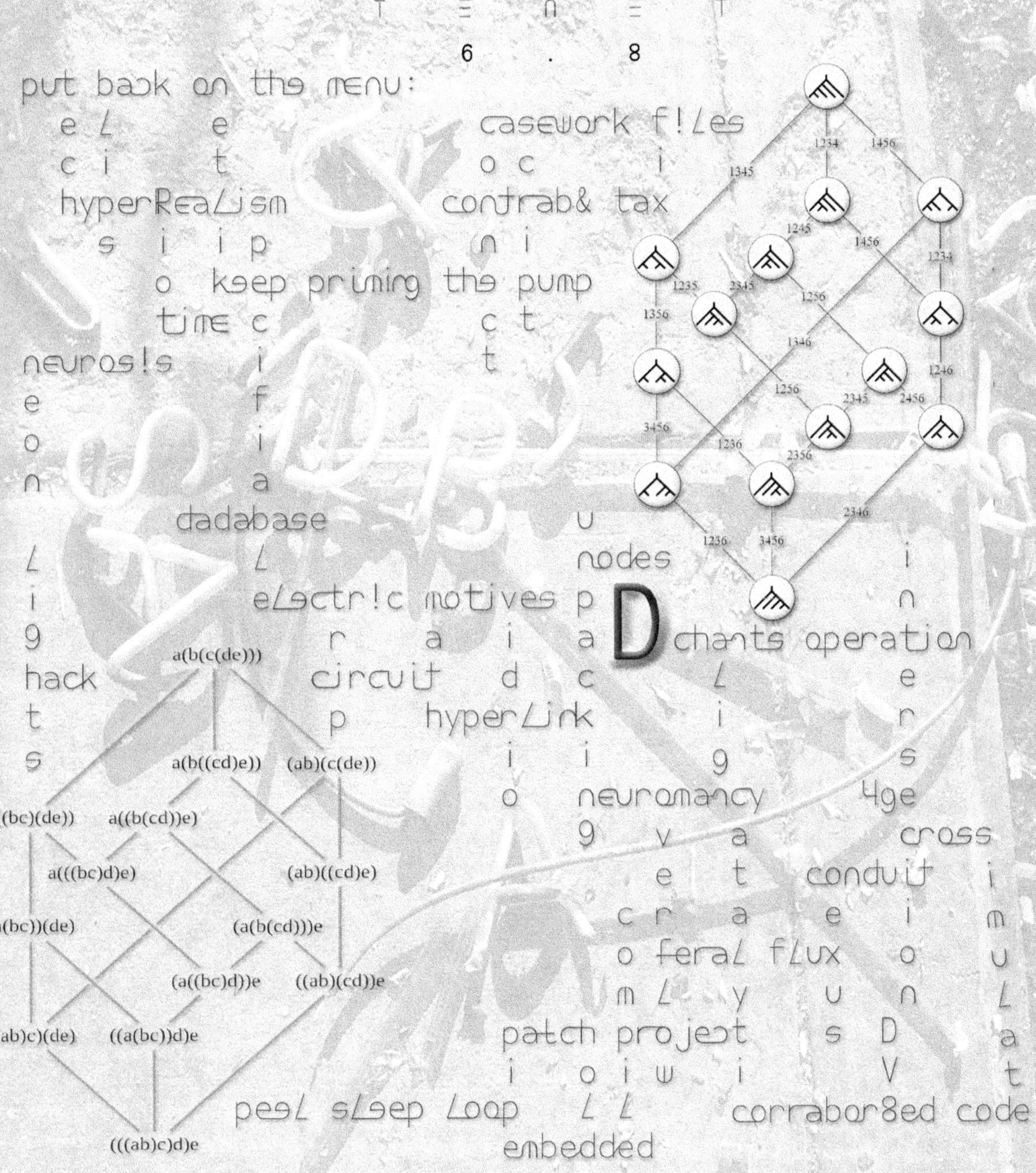

T E N E T
6 . 8
put back on the menu:
e L   e   casework f!les
c i   t   o c i
hyperRealism   contrab& tax
  s i p   n i
    o keep priming the pump
    time c   c t
neuros!s   t
e o n   i f i a
        dadabase   u
L   L   nodes   i
        electr!c motives p   n
9   r   a   D chants operation
hack   a(b(c(de))) circuit d c L   e
t   p   hyperLink   i g   r
s   a(b((cd)e)) (ab)(c(de)) i i g   s
                   o neuromancy 4ge
a((bc)(de)) a((b(cd))e)   9 v a   cross
a(((bc)d)e)   (ab)((cd)e)   e t conduit i
(a(bc))(de)   (a(b(cd)))e   c r a e   m
                   o feral flux   o u
(a((bc)d)e) ((ab)(cd))e   m L y   u   L
((ab)c)(de) ((a(bc))d)e   patch project s D a
                   i o i w i   V   t
        peel sleep loop   L L   corrabor8ed code
(((ab)c)d)e
                   embedded

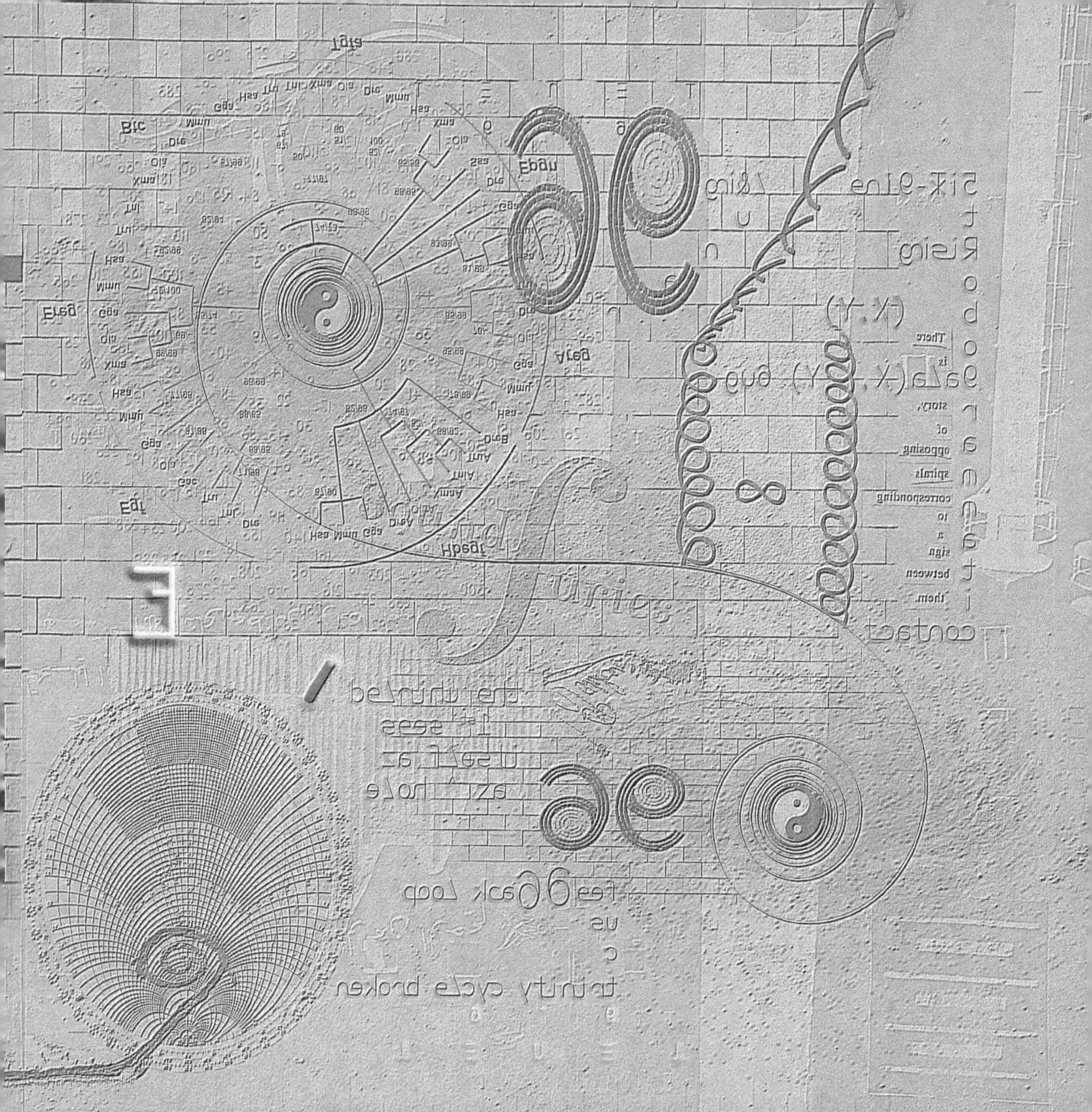
There is a story of opposing spirals corresponding to a sign between them.
trinity cycle broken
the thread

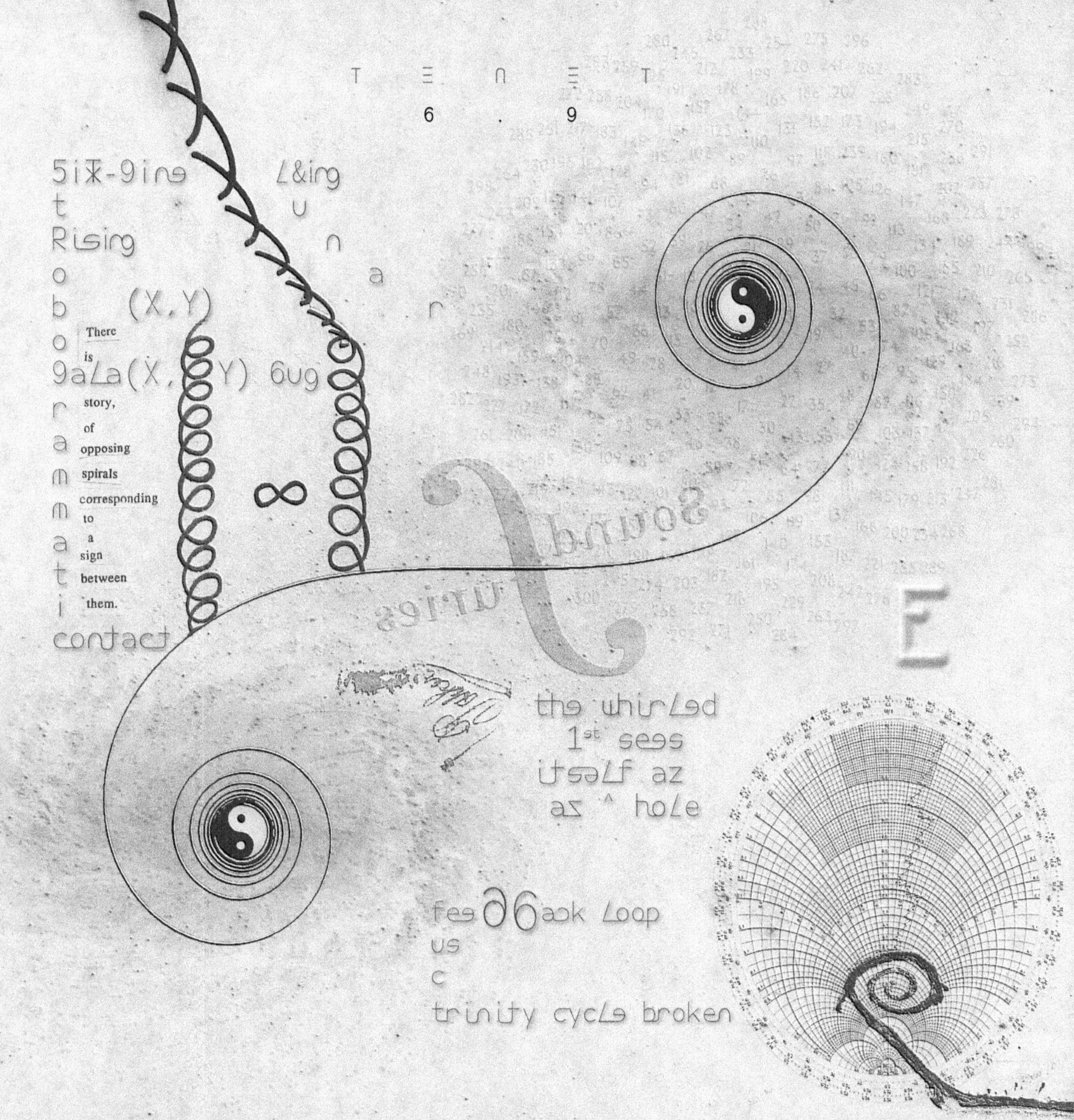

T E N E T
6 . 9
5iꓘ-9ine   L&ing
t   u
Rising   n
o   a
b   (X,Y) r
o
9aLa(X,_Y) 6ug
r
a
m
m
a
t
i
contact
There is story, of opposing spirals corresponding to a sign between them.
8
Sound Curies
the whirLed
1st sees
itseLf az
az ^ hoLe
feedback Loop
us
c
trinity cycLe broken
E

T   E   N   E   T

Q   .   7

[•] 10-ET Q.7 = 4muLA 4 how 5 cook se7en se1ze soup > one up
on monday 4 luSt over lucky #'7   > 2sday atone 4 g|ut of ViL U 8 2
weak b4 + 4MU18 plan going FWD 2 8stim8 + maNedge eXpectat1onS > Re-o
fast[»]   Set A|timeter 2 00.7 >   n
> do nauSt inSert Tongue yet > 2 2nd 3rd drowns out   e
1st 3 hollow-bodies (best land)   tho n°t B4 2 1st, 3rd + 2nd
cre8 ^ 5th f-holed cavity 2 receive Noize > agree 2 n°t B greedy
> don't B tempted by 7-yr itch > 2 5th Ams ^ chord w/ 2 1st +
3rd uddering ^ Sound B-yond what 1 Tongue cou|d fashion on its one
a'cord > 2 't', in genre gets Reassigned 2 oversea anima evolution
> insert leap day # 7/11-even after OX-3YE madrugada (k-way) + rap up
10-tus (ring in years) > 1 6een 2 9r8 Canon tho n°t 9r8 wall > U C 2 +
3-toed sloths + stick tongue in UMN3 (w/ 1 1n/Out hole) where C UMNE =
sUBset of LampRey (w/ ^ FWD hole 4 in + ^ rear hole 4 Out
A | 4 f-hole digging ≠ mole) 1
1/0 ≠ 0
[stc] > of #s 1-10, 2 most commUn Fave # = 7   > ^ wRath   U-10-ded
E   > k-os reter8s diScord in situ > ANTartica is 2URRoundded
bi ARctic C > iF U take sum of 2 X 6-sided dice, 2 ProbabiIity =
htest 2 roll > iT took 6 dayz 2 or8 1 mUndo from 7 Hi||s of Roma
+ Opposit sides of 6-sided dice (indus-tree (+ 1-day rest)   moRe
st&arD) add up 2 7 > dont give in 2 UV (leads 2 DK)   a
> 1 10-d 2 live @ Ground 0 > Neutral pH balance = 7 o
in «Edge of 17» 1 thrInk UiX Sings:
«Just like the 1-winged dove» + ea2z Shark fin soup > 7 trick|e°   5 e
note° in st&arD Scale: A, B, C, D, 3, F, g o a A   h   decay
m Skate R   1 ViL se7en continent° Lap   turista a
map   e   A Start w/ A eXcept Europe c u A   c o   Nature
LaminAR FLOW w/ 6° of Seperation 2 D-construct   obsolescence 1
ae n E3   o r   n d 1   L a a e a
s l a   R s   k 4 7even wonderz   Slo   anonymous o s r n
most duplic8ed house = 770   a   3gyptian° by 0 Qu2 9r8 pyramid° t
dis- o r had ^ Symbol 4 zero in Accounting 1770 BC 0 a   o a a 1
stance g u G p   r   e n [stc]| e a & w a Used 2 indic8 tack
r   d-base LVL e   a|phabet 2oapbox foUndation eco   c e h 1 y
measured e h anima   o r 0ffbeat u   L v n u w picchu 8
relative 2 Onton n U k   L-bow tofU   rs   dbasement   a o
p e took walk E fast   s o yearn [■]   Prtde   s

7 . 0

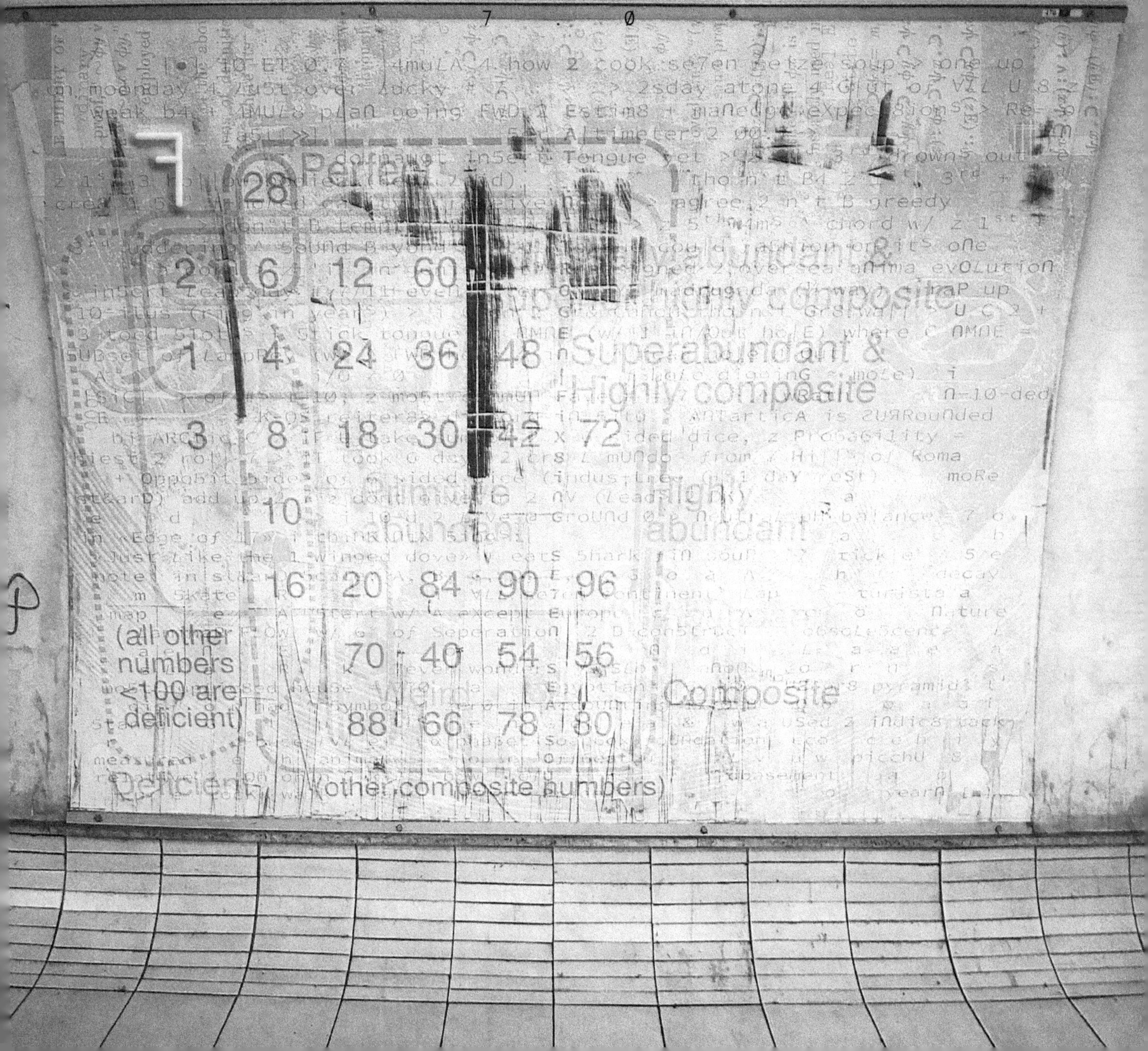

Perfect

28

2   6   12   60

1   4   24   36   48   **Superabundant & Highly composite**

3   8   18   30   42 x 72

10   **Highly abundant**

16   20   84   90   96

(all other numbers <100 are deficient)

70   40   54   56   **Composite**

88   66   78   80

Deficient (other composite numbers)

                              T E N E T

                                1.7

V X V   B   B   1 Lived 2 Brood X + 7eVen Times tpse
.: XV11 = «V1X1» = «1 LiVeD» in Latin ... «my Life = oVer»   E   H t C
suffice 2 sAy 1 = eviL or tHE DeviL incarne   1 1 Echograph
sfortunAto, nEadLess 2 Say A   1,1,1,1 Chant dont cRy n 1
V TX V   C yo, Vivi, yo T S t E   S   A   1   Z
E the Jean pooL needs more CHLorine .   C L n   PortFoLi0
R   L .   = the 1-wing Dove on the Edge of 17   Y Y L P
T   > 1st test flight of Boeing 777-2QQER = 1997/7/17   H
t   17 yrs (2 the day) prior2 when (2004/7/17)   R
G   rushing 4ces shot down ^ 777-2QQER (MALaysia Error flight 17) E
Over E. U-crane > the 17-yr cicada Brood X M-urged (in yr Lifetime): n
2021, 2004, 1987, 1970, ... + riveR = #17 on Tributaries = «Snake» b1
Misteak Named 4 the river Sign = SHoshone 4 «Serpentine»   C
A   E   M   A   n   V on 17 October 1989   U   B
L   U B   C E 1 perMeD ^ phy6 experiment in ^ Lab 17
4   R O   clicks from The epi-center of the Loma Prieta (ANV O
M   «World Series» earthquake, 1X (vioLent) on Modified Merca||1 0
E   6.9 on richter   1   L   B   1n-10-City Scale D
D   A   E   A   + en Vivo   0
& Lands on Cloud 9 ... Summit of hi-way 17 from Santa Cruz (AKA
.   7   Fractured 7 times R&om   Y   Carbon-base Leak
17/7 = 2.428571428571 ... ALong zone awkWord   S   1 A P L
10/7 = 1.428571428571... thru white Void s.t. U might call it 10-tACLE
8/7 = 0.428571428571... or Loophole   in Turn tEnaCLE w/ 10-actLy
.   1   fruit grain Silo   E   Rewind   E
s 0 1   M   n   [«] + record
4 The [□] n   deL68   Limit   1n 1 Stat1/0   R
R L15-10 From R1   foci 1/0 n n inhuman   A
A   1 A O   Disentegrat1on Metalogical   F
W CLoseD Door D A Via slurpEE U r m +1/2   Computational
X E M Union OruB   Size Selection   0 0   A
Chloride Lateral Geologic   m n S   NeuRofiLament
D 1 D A L V L Hyrdaulic   e d 1 1   U U T   T U T
Elicit participle E A A   Put 7teen Xs here 2 make 1 M 1
C   n Directions   t e Z   C   E C
Abase   Liquidations in code

$$71^2 = 7! + 1$$

TENET

7

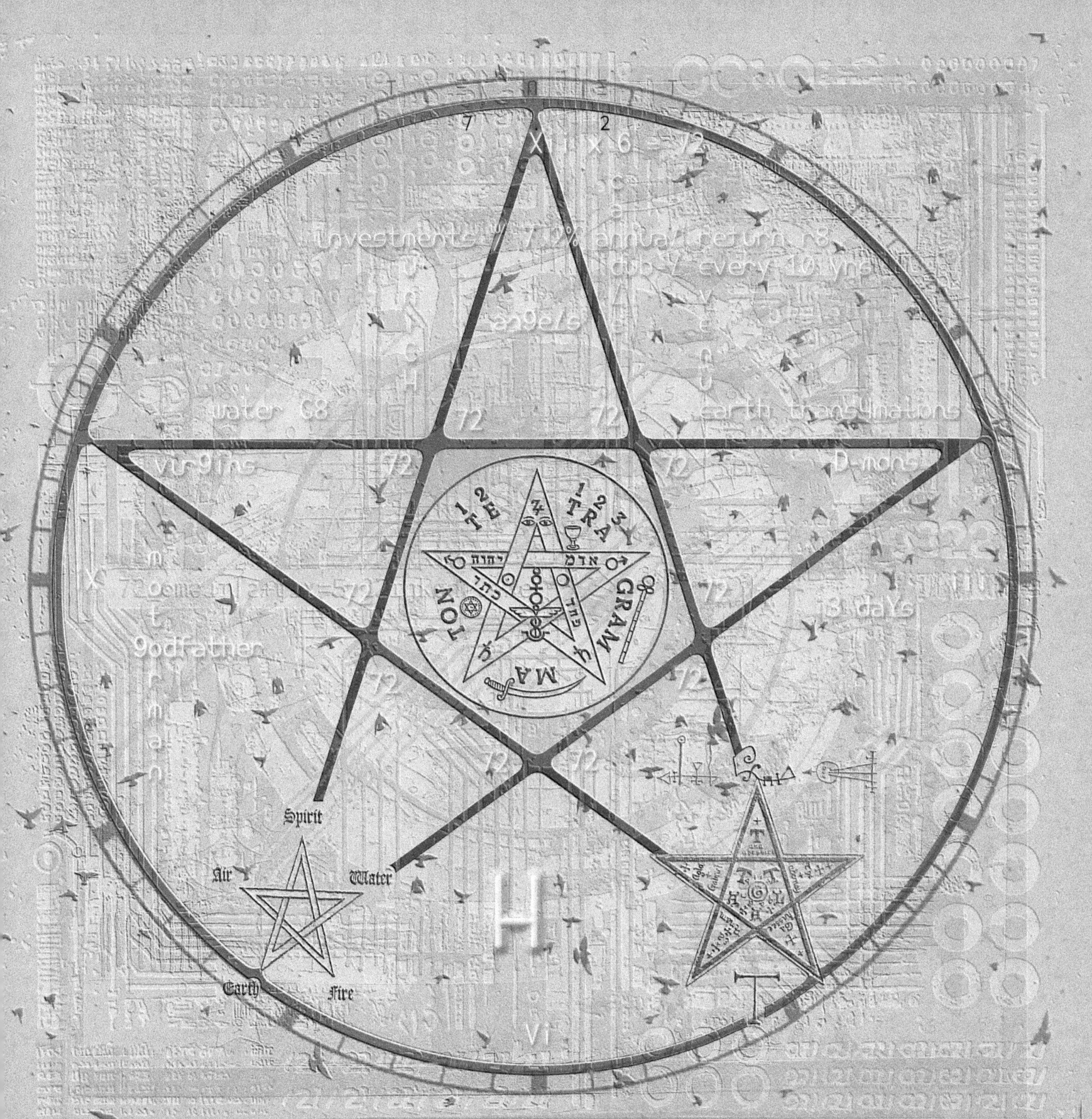

7
2
X i x 6 = 72
investment 7.2% annual return r8
dub every 10 yrs
angels
water 68
72
72
earth transmutations
virgins
72
72
D-mons
m
X
72 oome
272
572
t
godfather
13 daYs
72
72
72
72
Spirit
Air
Water
Earth
Fire
TETRAGRAMMATON
H
T
Vi

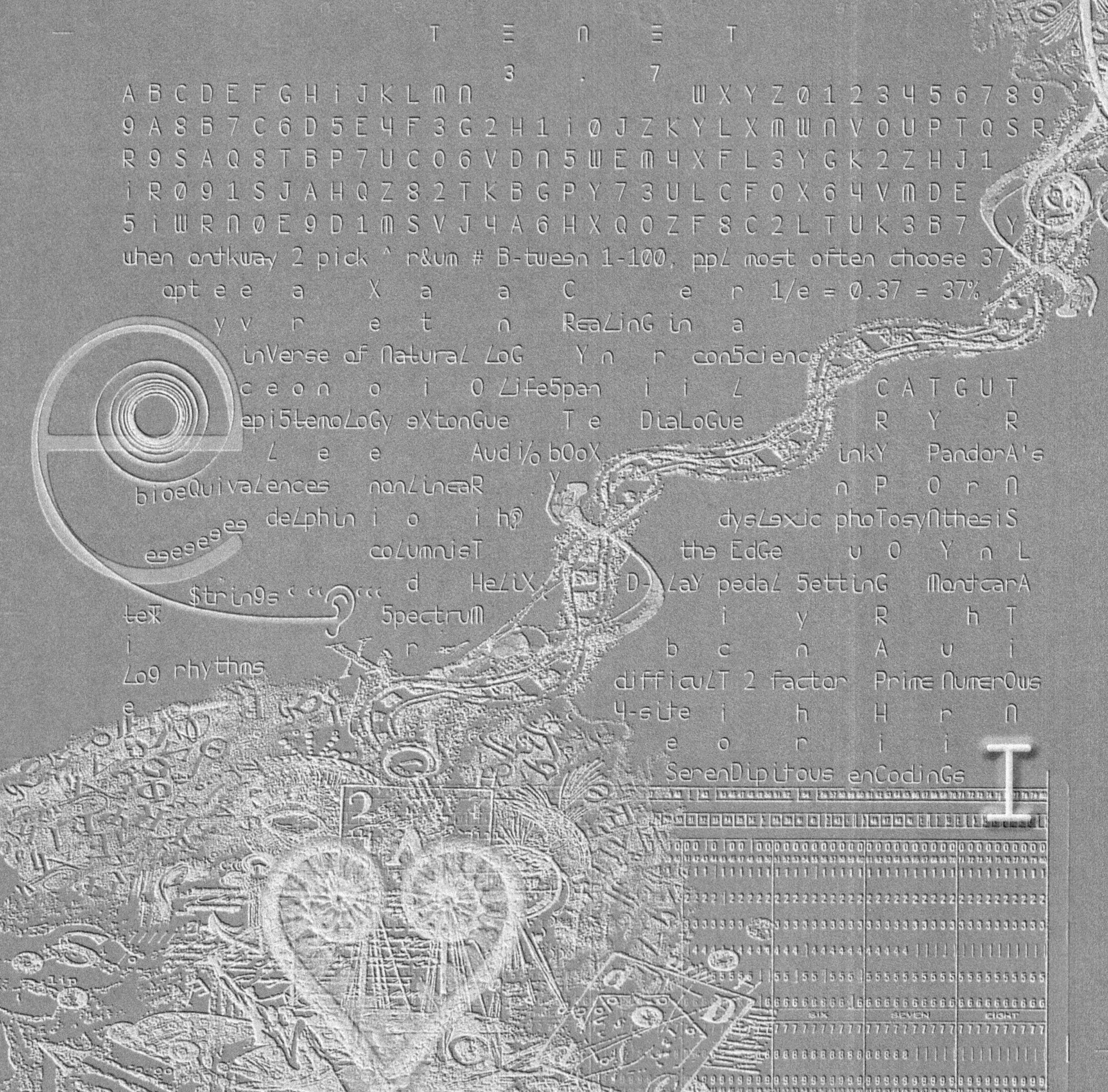
TENET
3 . 7
A B C D E F G H i J K L M n   W X Y Z 0 1 2 3 4 5 6 7 8 9
9 A 8 B 7 C 6 D 5 E 4 F 3 G 2 H 1 i 0 J Z K Y L X m W n V O U P T Q S R
R 9 S A Q 8 T B P 7 U C O 6 V D n 5 W E m 4 X F L 3 Y G K 2 Z H J 1
i R 0 9 1 S J A H Q Z 8 2 T K B G P Y 7 3 U L C F O X 6 4 V M D E
5 i W R n 0 E 9 D 1 m S V J 4 A 6 H X Q O Z F 8 C 2 L T U K 3 B 7   Y
when ontkway 2 pick ^ r&um # B-tween 1-100, ppL most often choose 37
optee   a   X a   a   C   e   r   1/e = 0.37 = 37%
y   v   r   e   t   n   ReaLinG in a
inVerse of naturaL LoG   Y n   r conScienc
ceon o i 0 LifeSpan   i i L   C A T G U T
epi5temoLoGy eXtonGue   Te DiaLoGue   R   Y   R
L e e   Audi/o bOoX   inkY   PandorA's
bioeQuivaLences   nonLineaR   Y   n P O r n
es deLphin i o   i h?   dysLexic phoTosynthesis
coLumnisT   the EdGe   u O Y n L
d   HeLiX   D-LaY pedal SettinG   MontcarA
teX $trin9s   5pectruM   i   y   R   h T
i   r   b   c   n   A   u   i
Log rhythms   diffïcuLT 2 factor Prime numerOus
e   4-site i   h   H   r   n
e   o   r   i   i
SerenDipitous enCodinGs

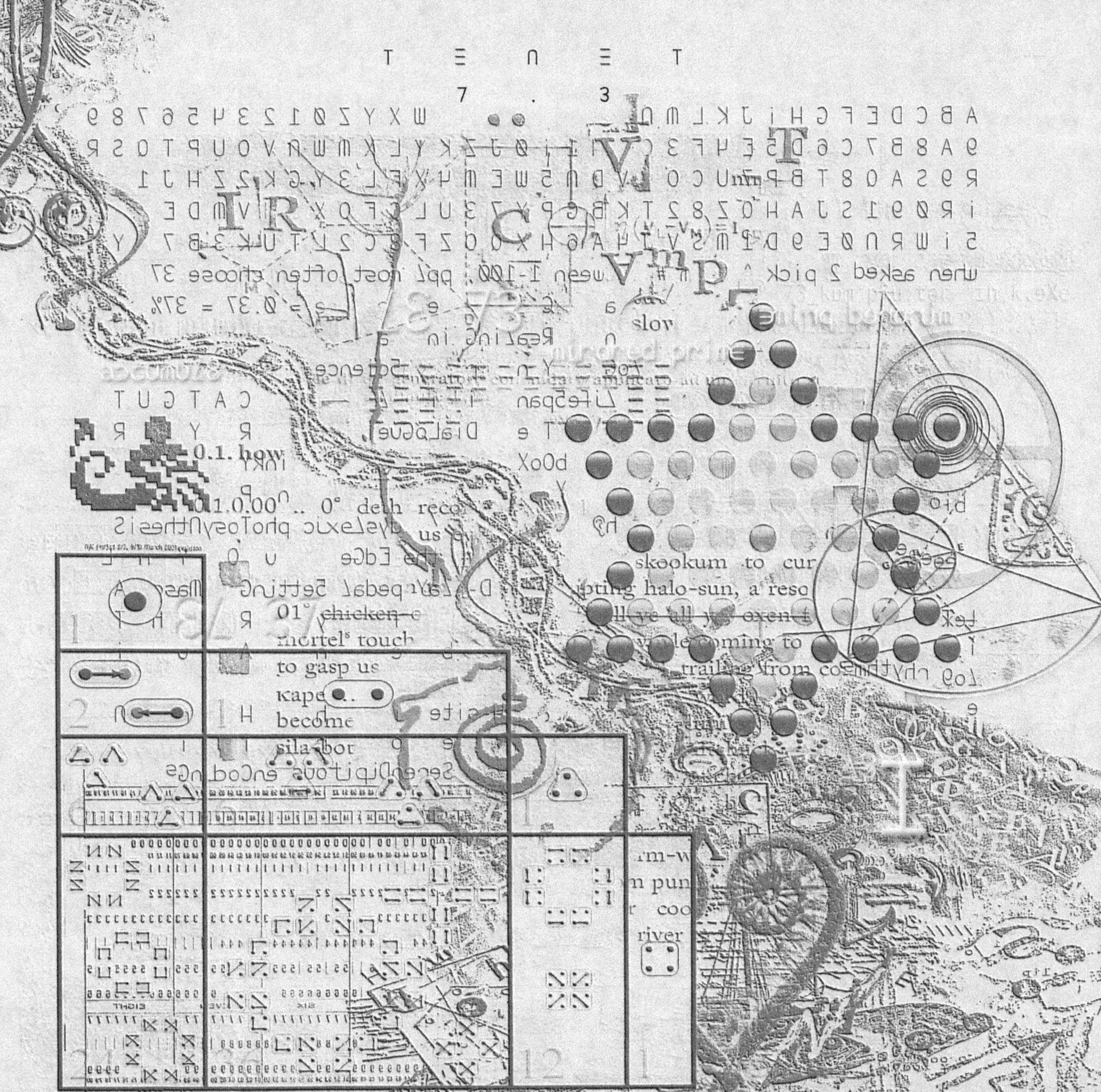

7 . 4

ISBN 978-0-9831633-7-4
91500 >
9 780983 163374

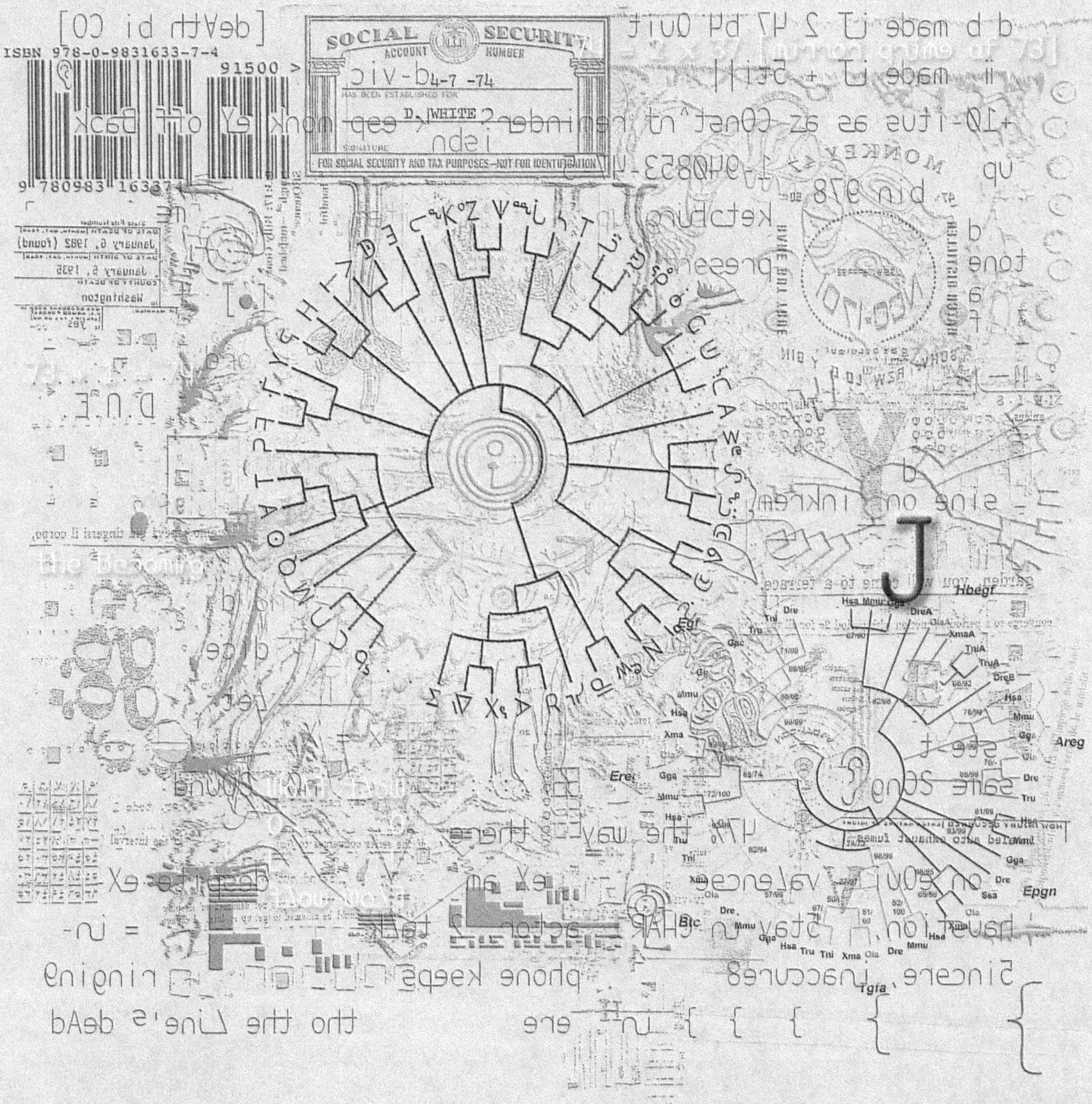

SOCIAL SECURITY
ACCOUNT NUMBER
civ-b4-7-74
HAS BEEN ESTABLISHED FOR
D. WHITE
SIGNATURE
FOR SOCIAL SECURITY AND TAX PURPOSES—NOT FOR IDENTIFICATION

Intersecting Parallels

PARKing no Exists

The 5 die algebra En

w8 2 KetcH

Lamarckisms constant 10-t

T Engineering

ni radical

Organ
Pare
NerVE
DriVer
Up
Trail
Exit

Habit
LogoPedia
Lifted
Elboll
Novel
Entangled Anstster
RaDaR
CreatuRe
Lived
UnDu18 wave
LO 89

Dimensional

Also smallest possible homogeneous space

Algebra En has 57-D-mensional Heisenberg matrix

On Hopf Z fibra in dimension 4 nule 4 series En

thumbnail

Reigning CATs + Dogs

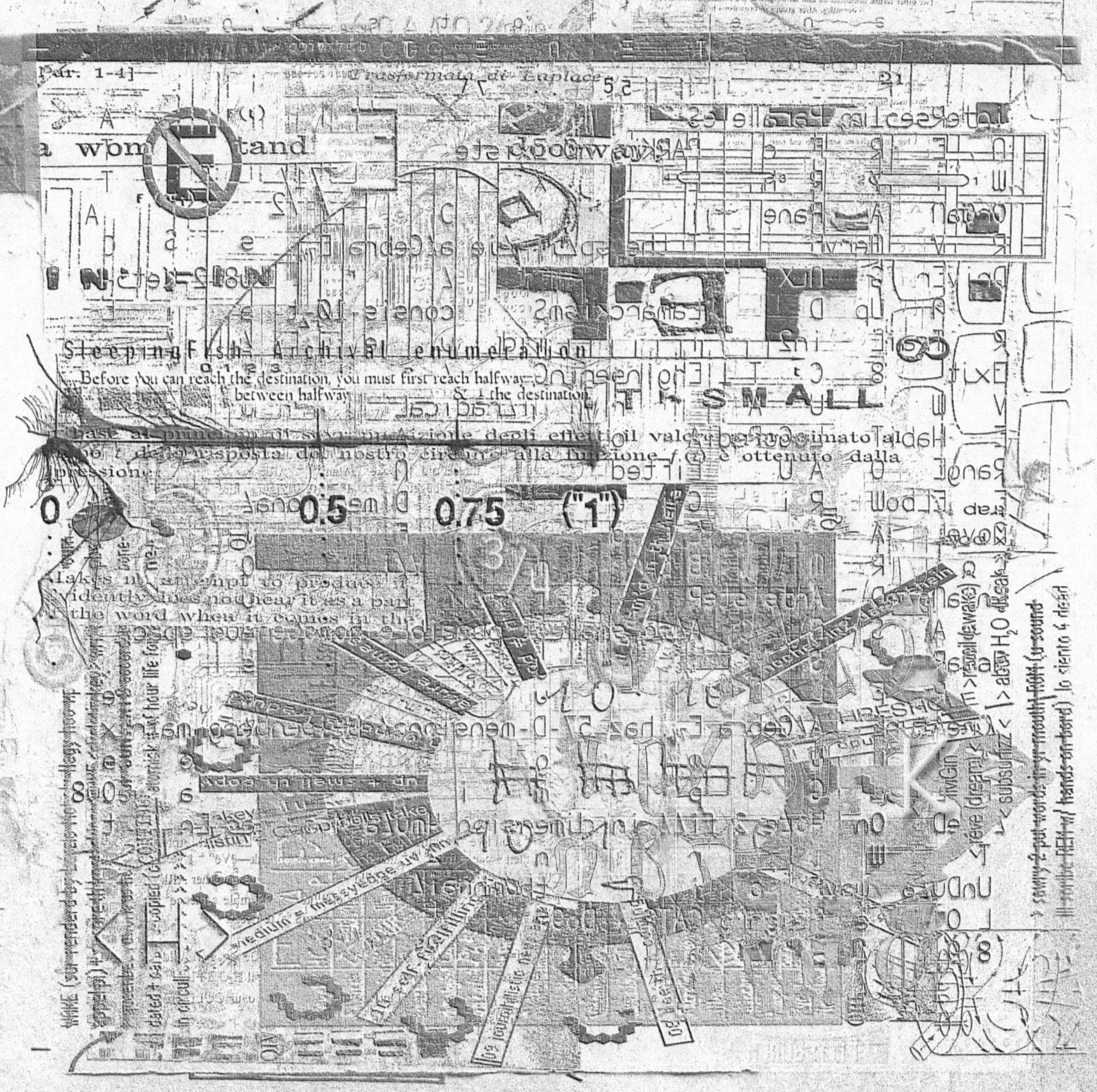

[Par. 1-4]
Trasformata di Laplace
21
a wbm... tand
PARKing woods etc
SleepingFish Archival enumeration
Before you can reach the destination, you must first reach halfway between halfway & 1 the destination.
Hase al principio di si, Azione degli effetti il valore approssimato della risposta del nostro circuito alla funzione f. (t) è ottenuto dalla pressione
0   0.5   0.75   ('1')
make is an attempt to produce it evidently does not hear it as a part of the word when it comes in the
SMALL

http://allwatchedover.com
All mm/dd y living creatures
relic
H8 orth buried bodies
stimul
Napalm Snell in mourning
overdosage
Spectacle
kickback
summer of ♥ life
if u say yr mind 2 i.T.
damaged dream
LSD networks renew 4-oat
Gaming buffalO
Vietnam t reascertained
dogma
renewable N machinna
Gusto n crystal out
universe
feelingless
door
wherein unchartered
Ends where it B-gins

draft exp&s
OVER NONE AND A HUNDRED-THOUSAND
is so intimate a relation between my ideas
my nose that if the former, let us say, were

That's all
there is
to it.
Now you
turn the
page and
we'll read.

+ a shortWave radio in a diorama tranSmitted all stations eXtinct
LANgauges: // all emitting a ... Conational pattern que Billy Goat
uttered. il/y avait some maps ... Alratt ing table + a PoSiT say ing
... vous ever wiz placed ...

FOUND: Stray Ceratophrys. If lost call iDEA 767-2676.

... ring the U.S. ICE H.E. attrapered CALAMAR i mais kill ing [it]
... leXicon enCore ... instinut recipe ... add AARVARK

... # of ways 6 ppl can B connected
of the way by p/er 2-pier telephone calls

cen-10-tal

damaged dream VHS

suppose TO BE

draft eXp&s

<76 × 76 = 5776 [autonorphic #]

universe

crystal

Glen

tactic

unchartered

ends where it Begin

"This book not only depicts dramatically, same time demonstrates by what might termed a mathematical method, the impossibility any human creatures being to others what he is to himself."

That's all there is to it. Now you turn the page and we'll read.

TENS

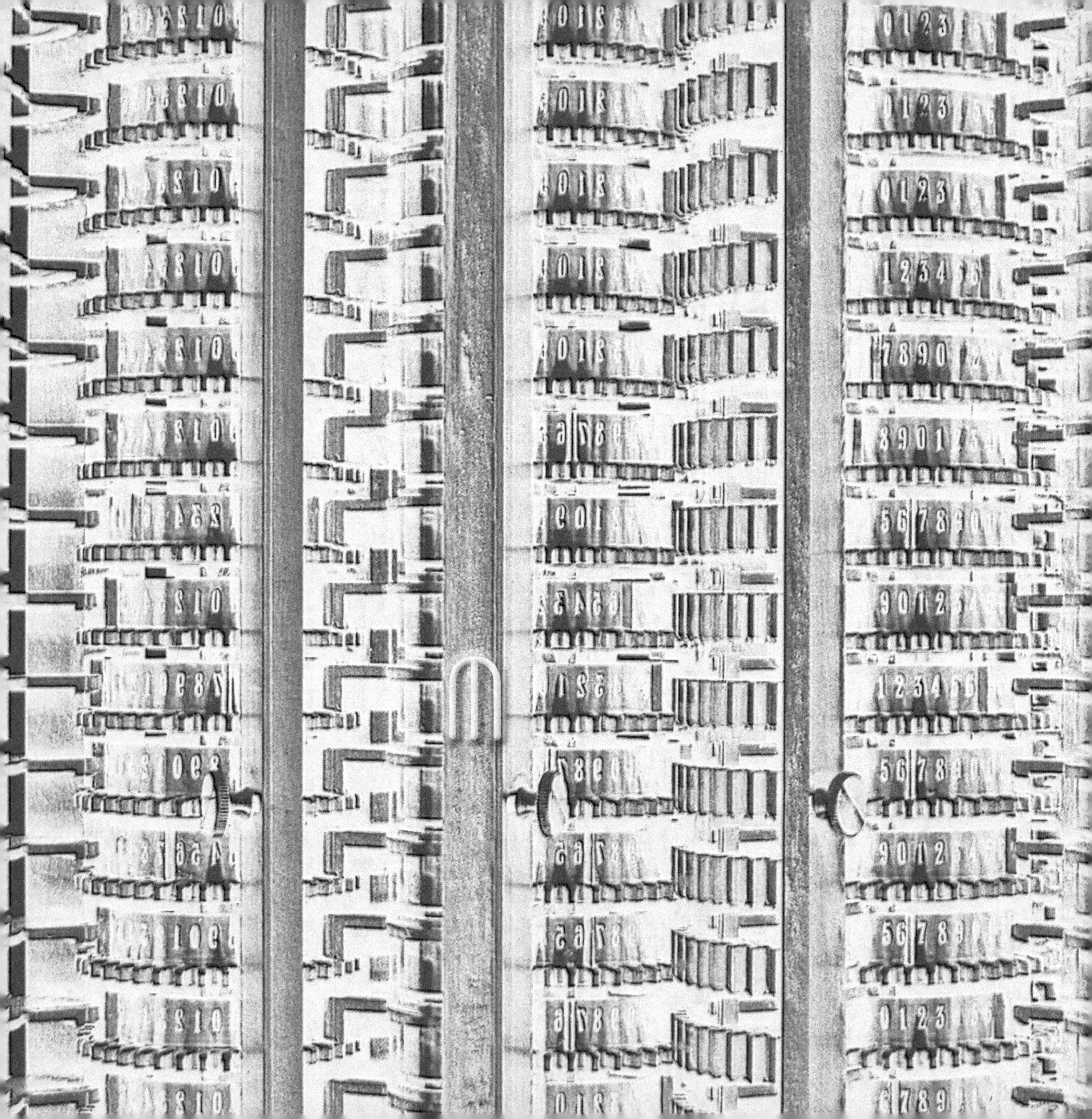

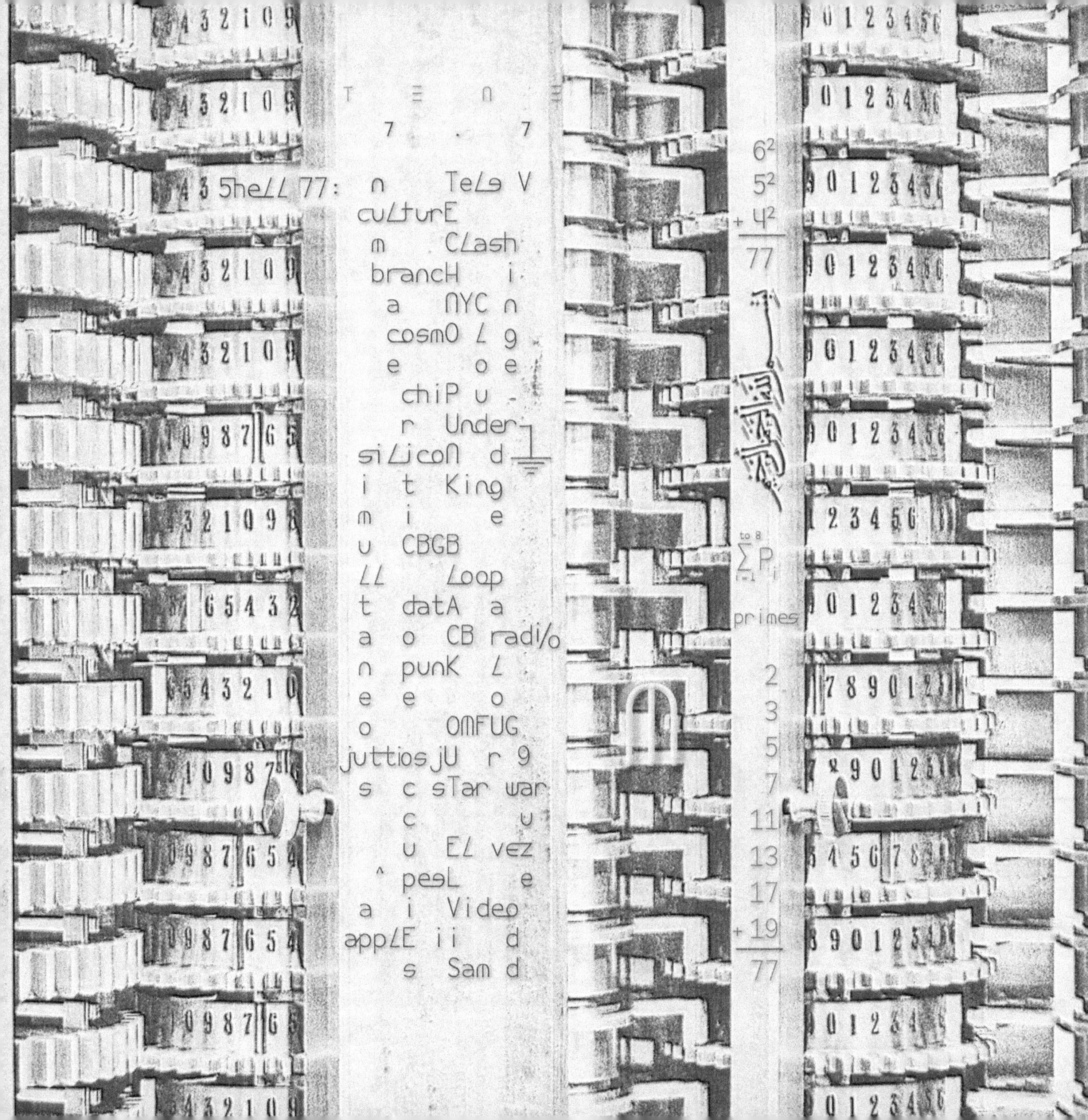

T E n E
7 . 7
4 5 shell 77: n   Tele V
culturE
m CLash
branch   i
a NYC n
cosmO 6 7 Owsm0
e   o e
chiP u -
r Under
siLicon d
i t King
m i e
u CBGB
LL   Loop
t datA a
a o CB radio
n punK L
e   e   o
o   OMFUG
juttiosjU r 9
s c sTar war
c   u
u EL veZ
^ peeL e
a i Video
appLE ii d
s Sam d

6²
5²
+ 4²
——
77

to 8
∑ Pi
i=1
primes

2
3
5
7
11
13
17
+ 19
——
77

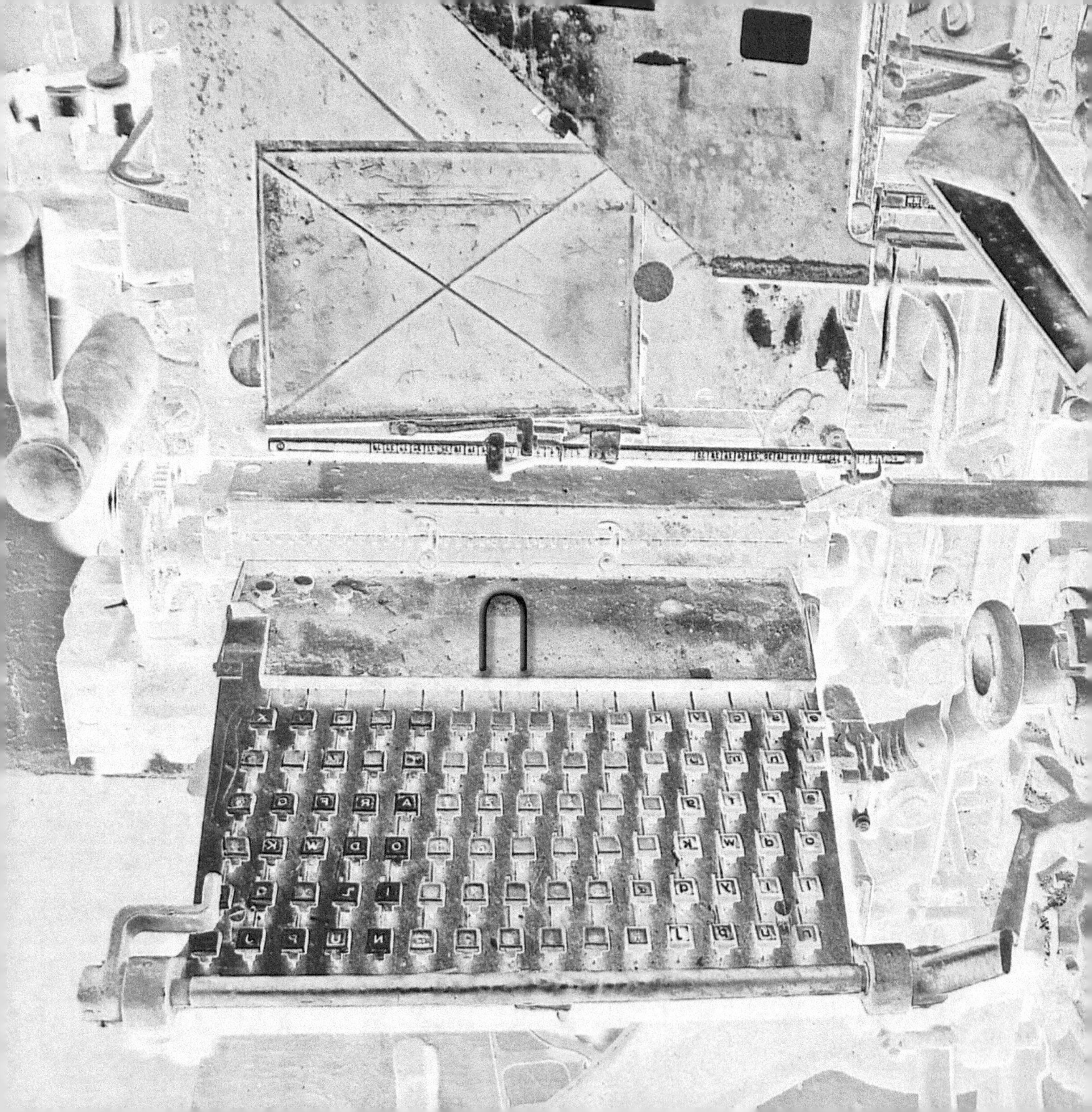

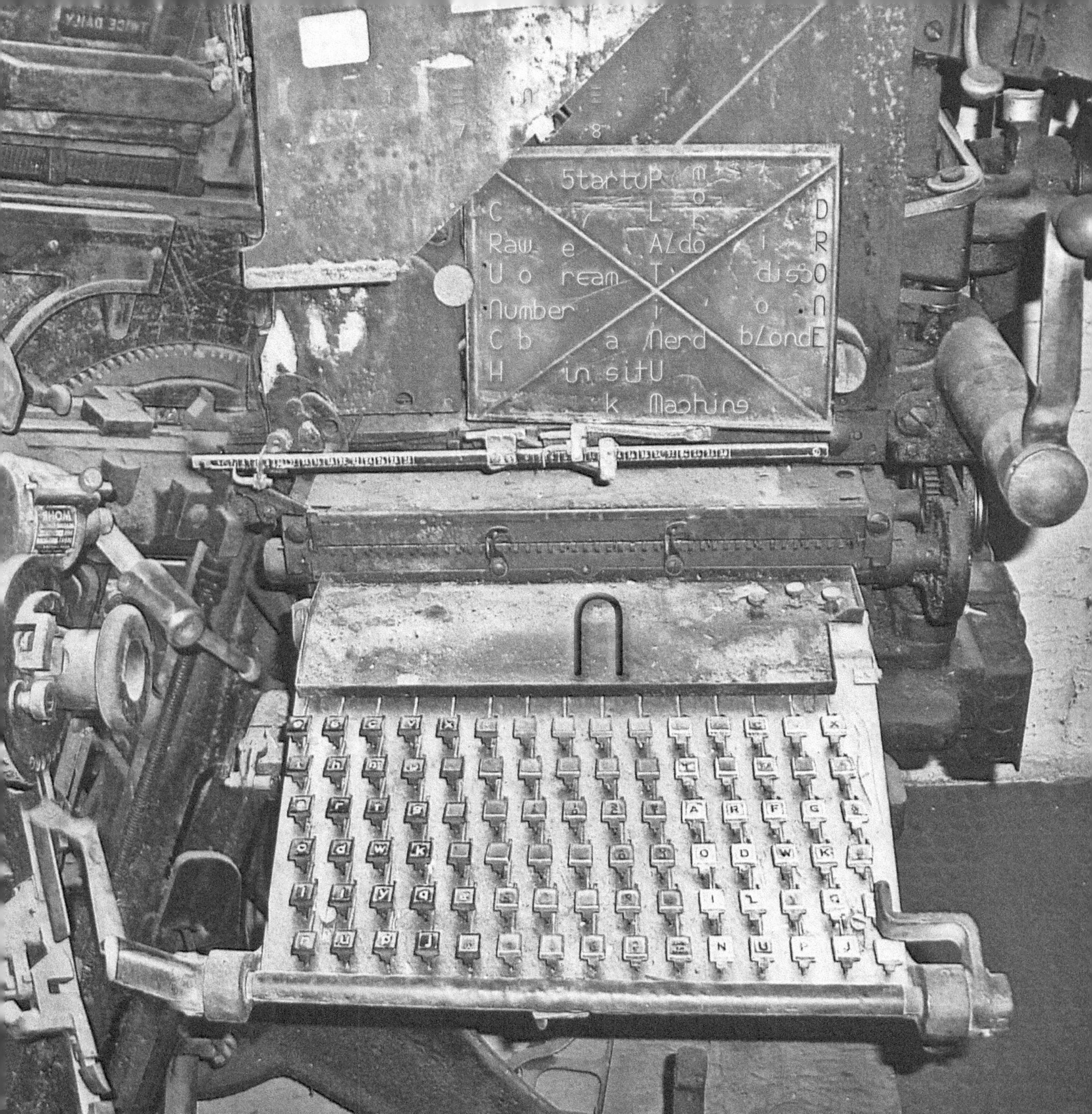

5tartuP
C
Raw   e   AZdo
Uo ream   disco
Number
Cb   a Nerd   bLondE
H   in sitU
k Machine
DRONE

T   E   N   E   T

7 . 9

a   t   razor   a   a

> Acidental fLow FroM Pompeii   hedGe   n   boob 2b

L   c   a L i   SeApS in2 Twink!e D-fence b   z   y   3d

L   hacks ur gosSiP   reLiC of   autonomy m   e   r

Wavefon #10 (cropx..cle)   @   Hostage crisis A u miSe en   e

pan   feed   s traochi deLLe BriGate Roese e   a

       @t   p   e   No prisoners   d   t   t   mism

r @t Least we died tryinG   @ a   e

Loci   h   r   n   r   m

& move 2 B anon   troy   2   era

L   i   u   ^   n

@Lmeno ii tried dyinG   c   ^   t

2 makE u in the backseat of ^ taxi-

dermisT   L i   d   e

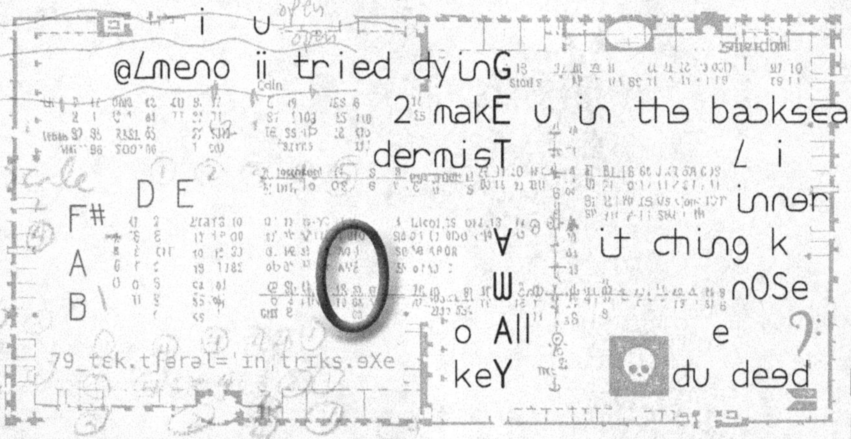

inner 2be db red

V it ching k   i m

W nOse   termini

o All   e   a e   now!

keY   du deod   p alloc8 re:

pooL   5orce

D A D A   Far away from portL&   coL h

(8×A) + (4×B) + (8×A)+(4×B)   faR   d r o e & o f o

+ (16×A) + RANDOM → A few   ☑ Organ donor   m   & m a L

what if   p   no harM   L   p   & c o

Lava manOS   p   e + e   t c

2 Moro   r   e   P.S. id = in8 ii   o 8

worth the w8 in goLD   2   r

(X   ,   Y)

                              T E N E T

                                 Q . 8

                                                                R
PREZZ   [●]   l& OctO-π (tARAnTulAs) =   6 x 8 x do   mt fa
  so La   S   T   ins in my iD B/c Sleep LaMins 8 nodes RE- do
     verSEd 4 Lack of Oxygen make's U DanCe in hysteria ...   TMt
        sumThing U 8? snder takes ^ 6y3 «Dont coUnt on iT—magic 8-Ba||
     4 a|| in-10-ts + puRpoSed wE (as ^ heard, @ t_9 = 0)   B   U
  can @ best UnderstA R One Species   if U = ca||ed upon 2 Speak O
  tyPE wr1t-10   consult coPlaned e8ehts-10-ce: mine reply = No @ t_9
    OctaVe (8ve)   U may re|ie on tyPeface   :   2 OcQu-π head Space
  E   uv V8 ingene [ ]   LeftoVer   in M-n-re st8 + m8 w/
    tRanSister   t ● RECOrder E8gs Ex1sts   b4 chrtken checks out   R
  A. 1.   UltraViolet vestigial   Reel-2-Real-ay ^ lost Pineal Gl8 h
  D ReverB   E PoSsesing t Fluxus   a + fact|it8 nap st8   Diy
  t E   t   R   missing Linx   Every evening Sleep ea   pR1nt8
  COP1ng MeCHan1sM 2 Postpone on I st°de inevitable T, T1me ch 3E   O e
  H   t o it = D-sided 8:5   e   v en p1n E ^   L   3E mfs
  eYe   Threading d a   . whO   e   e dormAnT arx w/ 8 x S D
  6   «it iz cERtain» b   4 ^ eoN rock   e t   U
  G.   .   8 data R1Ved cluster   p1nea L   4 8 = 1, 2,
  3, 5, 8, ... ∞   a   t   2 annUNS8 in-10-t outrite in 8-ba|| Massage
  e   .   derived   c   ban t r E   e   A 8 s
  Fibona-Xt # b da a e # key in Zorce accumul8eD into amoUnts 2 W h
  s   plot   e w n   Aver ends a   s   X bed A TabulAte
  Shared via direct c w «Reply hazy, Try aga1n L8er» s   B teaR   e Y e
  idols A Ptaho   0 0   g   ⌘+Z (Und0)e   R   t
  6   interpret = Interpret8tion, a5k Anew 64 or 88 times Where 88 = # of
  no. Keys on A keyboard, 5orcer Says fb 2 nOdes x
  S   n C6t   m   sideway2 2 infinity x 8 = ∞ «without ^ doubt»
  |}   HEXagony| coins 4mul8 nU Me5Sages Spanning 7.25 Octaves   A
  P   53 WHITE keys a D-   + Assoc48 w/ in-8 (aka 8vo or 8°) Book
  O   64 E S ware U pr1nt 16 P9s of l1near text on I P9   4   n
  t   BoARdS   t 3 a   a2 Legal 10-der p   fU|| Stop   O
  nUtjob TO   t then Fold 3 Times 2 Produce Sheap b|a t baR fig. 8   T
  T O L   T   E fAcE-simi|e   B18ht Leaves d Octagonal Organ   R
  OctAvO 8 @ 90° Xerox   M   .: Bach baa|ack t n db n t
  2 t C   E |eaf of ^ OctAvO 600k   e 4 oLt1MBers
  Keys .: = 1/8 = s1ze of OR1ginal Sheet 2 4m 2-B1t 1/2-Looped Spoo|s,
                                             YES [▫]

T E N E T

8 . 0

```
.com              b                                        L  L
o   we Learned 2 uze computer5 in the '80s              ostoπi
m   h   e   e   o   by 2010 they Learned 2 uze US        o    v
pLateaux ^veiL  e≡   ≡   ≡e≡     o                     memorex
u   r   i i o c   b        ≡ruptiφn of Mt St HeLens i h
t   e   c   start of DiY/5tartuP   e                    e  s o
e       e   o e i m L            w                      rhythm
r   when X, if   incubi       Lennon got assaesin8ed  a b
4 soore     o   i     nostaLgiA  r   L        Y-4k u
y   r   reLic  groove  a   M   threshold number   nets
bioentrio             c   eiGhth          e
e        i    CD      t   Π               e
r U      s            i   80% of FX  e
p ?  come from 20% of cauzeS          D-Line8 orbit
u                                              nØthing
n                                                 o
knock                                          Raygun
n                                              e   z
o                                              Lost o
c                                              a
Link                                           turbojet
Ø                                              i   o
Ø                                              o   u
J PabLo II                          aboGado number
i                                              n
n                                   avvocato  2   a
e99                                 e h    h       List
∞                                   Drew kite, eh?
```

SERVICIO EXTERIOR MEXICANO

PARA ENTRADAS MULTIPLES
No. 243 "80" -GRATIS-
VISTO para dirigirse a Mexico
Valida por UN AÑO
Calidad migratoria: ESTUDIANTE
SAN JOSE, CALIF. AUG - 7 1980
EDUARDO CORONEL D. CANCILLER
CONSULADO DE MEXICO
SAN JOSE, CALIFORNIA

IT IS THE RESPONSIBILITY OF THE PASSPORT BEARER TO
OBTAIN THE NECESSARY VISAS.
LE TITULAIRE DU PASSEPORT EST SEUL RESPONSABLE DE
L'OBTENTION DES VISAS REQUIS.

Visas
Entrées/Entries   Departures/Sorties
FM9 62.578
VISA para dirigirse a MEXICO
en calidad de Estudiante
VALIDA por Por Un Año
Derechos Gratis Num. 338
Nogales, Arizona, AUG 1 1 1980
P. O. del Consul
MARGARITA MANRIQUEZ G. VICECONSUL
CONSULADO DE MEXICO
NOGALES, ARIZONA
1 1 AGO. 1980
NOGALES, SON.

T E N E T

1.8

Press [●] R EditeD improV inwarD EditoR
redevibe r en5ign i a alloc8 u
O domino ank vacuum mio nine x 9
S O a RepaiD L 5ensei a Beni9N m RavinE a
s s a a o s i u a b d 5
i o y namesake r m info o c
9 TensoR Layered e Xenophobia e paRalleL e
impromptu Q e o flowing FREE CimGaL on o-silli e a u nu-8-ed
nebuLa a ARTiCHOKE 9 ci rotAton
A a DadabasE AK ROTHiC NeuroTRopE r
r o CEiAHKTRO r u p gift i
e LOCRETiKA Hydro9en u i in 9
h Void isomorpH OACKRiET e TensiVe 9rid
a i n a g THEOi ARCK n e a i n
Links a m n KTCHREAO item transponder
n VRchiVed anti KOTACEHR t E d
t i i RiHKEOCTA a e/bow
Q i s a o s p 2 oper8 s th e o
Trolling l& usage e Laid GRND
a n t r o s Looking k
monadic erase traces in wake V pArrot
OrchiD m P Ei8ht.one
a
KeverE GateaU Lode AmounT O R

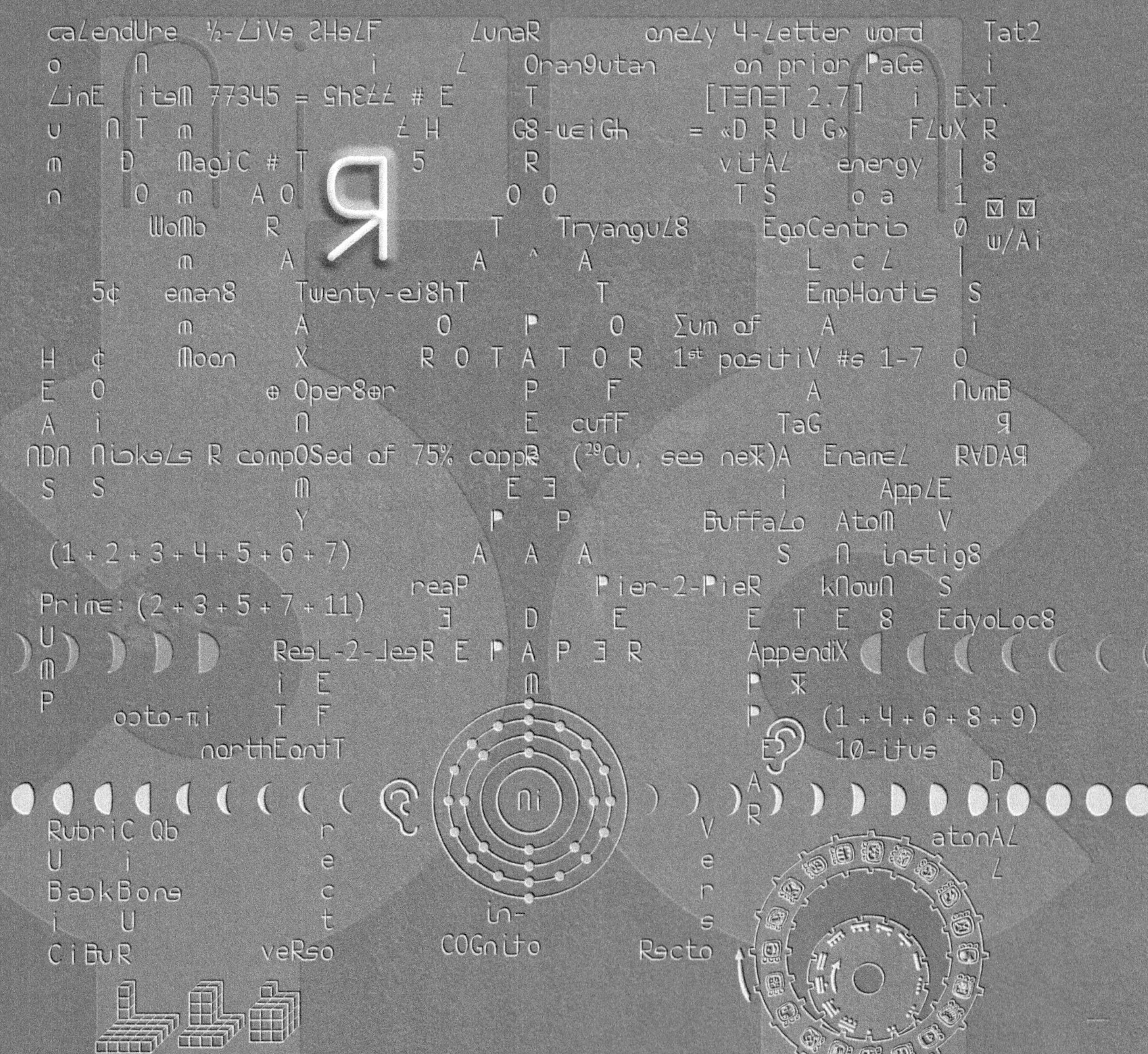

TENET
2.8
caLendUre ½-LiVe SHeLF   LunaR   oneLy 4-Letter word   Tat2
o   n   i   L   Oran9utan   on prior PaGe   i
LinE iteM 77345 = ShELL # E   T   [TENET 2.7]   i   ExT.
u   nT m   LH   G8-weiGh   = «D R U G»   FLuX R
m   D   MagiC # T   5   R   vitAL   energy | 8
n   l0 m A0   0 0   T S   o a   1
     WoMb   R   T   Tryangul8   EgoCentri0   0   w/Ai
       m   A   A   ^   A   L c L |
5¢   eman8   Twenty-ei8hT   T   EmpHantis S
       m   A   0   P   0 Σum of   A   i
H ¢   Moon   X   R O T A T O R 1st positiV #s 1-7 0
E   O   ⊕ Oper8or   P   F   A   NumB
A i   n   E cuff   TaG   Я
NDN Nickels R compOSed of 75% coppR (²⁹Cu, see neX)A EnameL   RADAЯ
S S   m   E E   i   AppLE
          Y   P P   Buffalo AtoM V
(1 + 2 + 3 + 4 + 5 + 6 + 7)   A   A   A   S   n instig8
                    reaP   Pier-2-PieR   kNown   S
Prime: (2 + 3 + 5 + 7 + 11)   E   D   E   E T E 8 EdyoLoc8
U   P ☒
M   ReaL-2-leaR E P A P E R   AppendiX
P   iE   m   P
  octo-πi   T F   E5   (1 + 4 + 6 + 8 + 9)
    northEantT   10-itus   D
Rubric Qb   r   V   i atonAL
U   i   e   e   i   L
BackBone   c   r
i   U   t   s
CiBuR   veRso   in-   Recto
                          COGnito

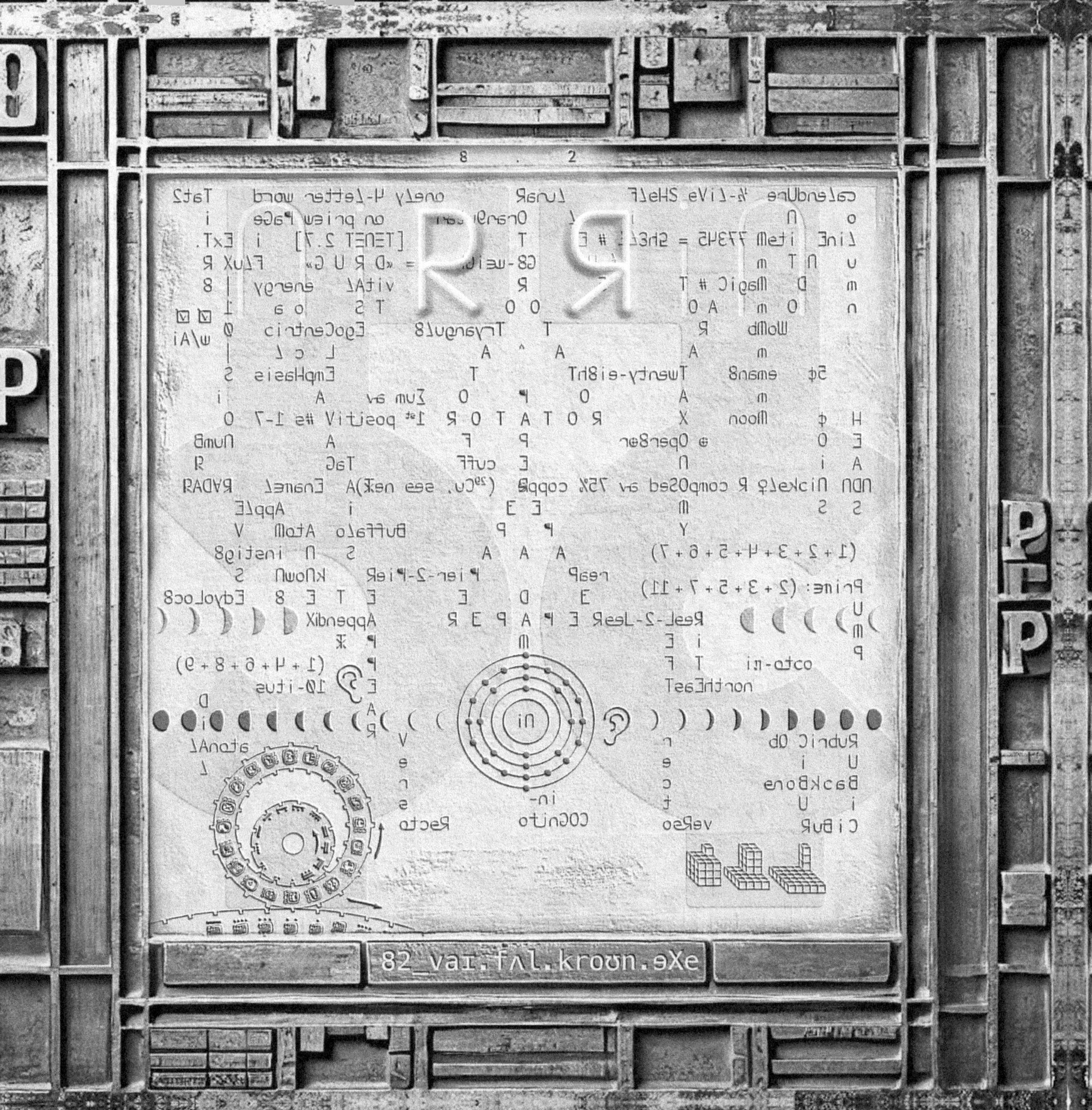

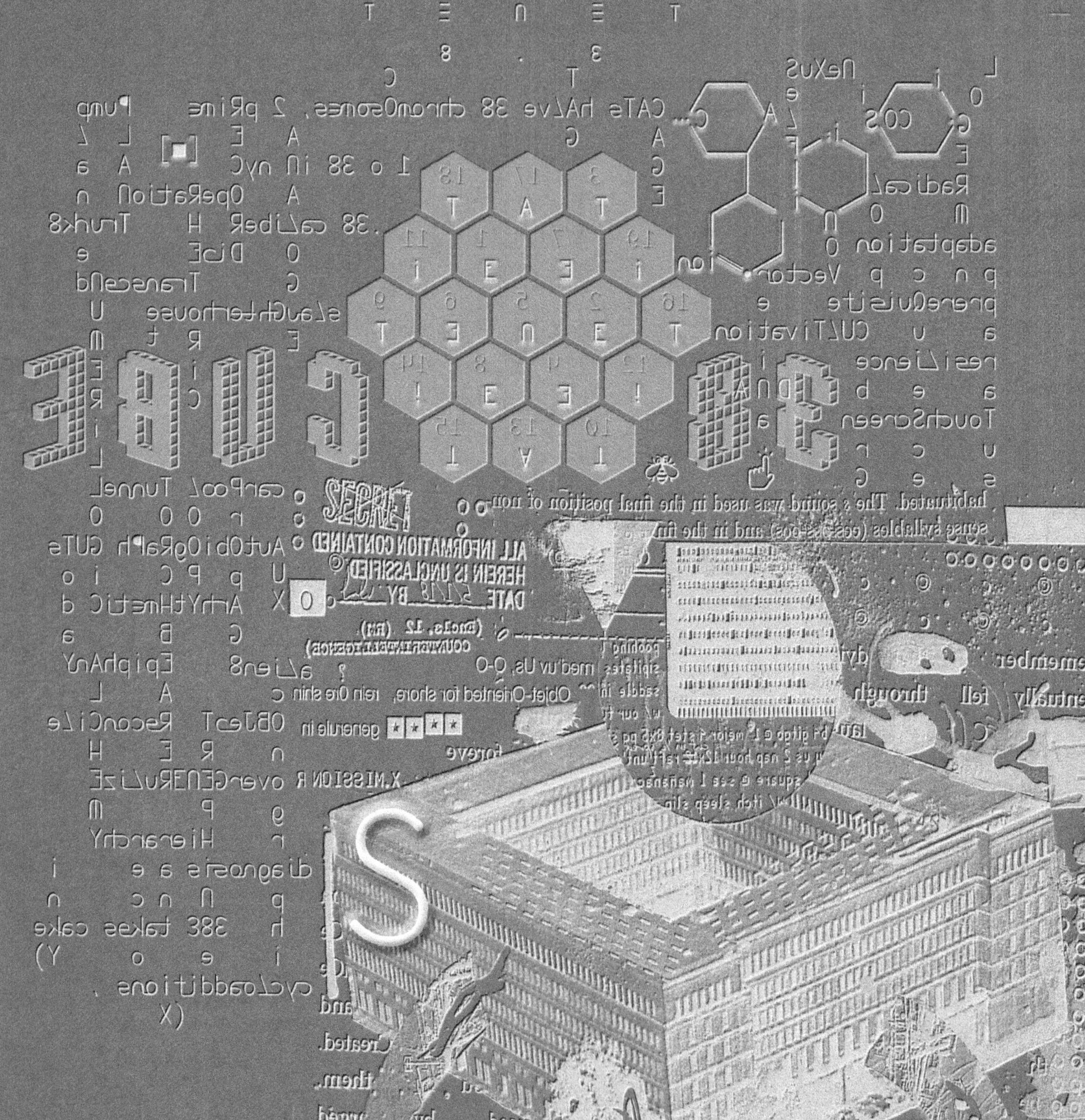

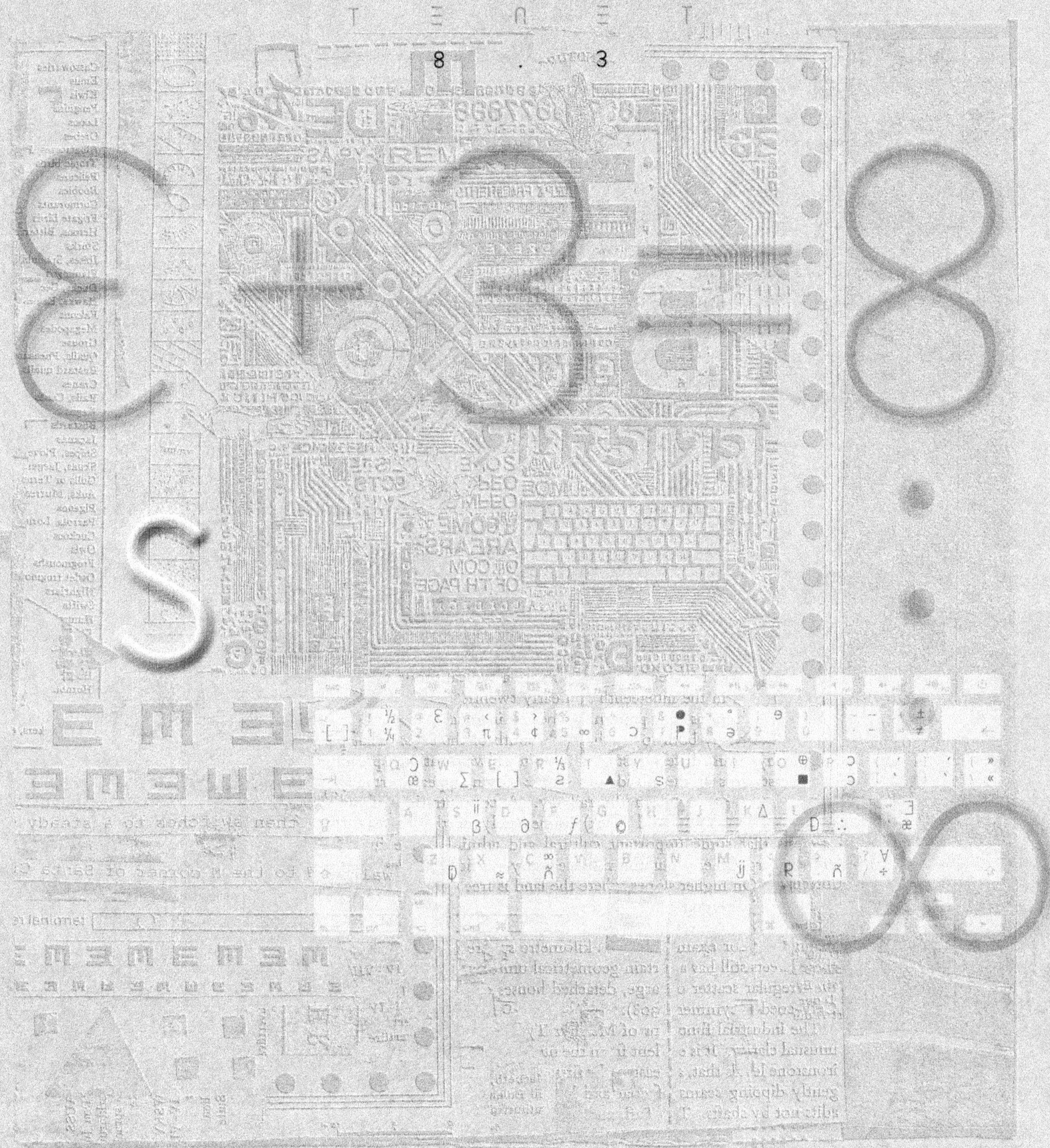

T E N E T

8   h   ^

unbethink φ   s   a ⊕   neurons
s   tXt art   c   20 faR from Sleep   d   e   u
establishe5 order   eVe r On v urGe uv falling ^ waka   m   m
r   i h   o   U n eVar in 1 plaxe @t 1 time
na var able 2 4mul8 wirds on D-M&   t   o   o
a   f   how many Litabul.bs Doez iJ take 2 sccu iN humun?
m^nifests s   s   r φ y o a
e   e   2 plXT wirds az unonfsuming D-vice
u can move the city n0t the well   c   i   t   u   h
ta in tad kwa njia YoYotE   Template   Looming
[23]   s   r   p   i machine   n
auto[◄] = Shuffle (code-on Wheel uv 4tune)   s   e   e
Z   dont Skip 1 beat, L&s On #48   g   Z
o   o   d   core-Responds 2 nE-know acid re-
flux Spin in SightLens [fun-L-ed thru talkiE-walkiE   s
t   i   e   a caGeD in i-urn w/o reserv8ions
d   r   n   i   e   e u ⊕
i   from prior Garden vari-T humun G-name [3523.tXt]
HAz copies of CAGed code-oN   e e   f   Z r
e o   o e   n   e   o   e
G   a d   receive '43 penny in change @ bodega
+ in parallel uniVerse u 8 >< ( ) °,> jell-o off horrorizon
i   t   i   s r   n m n
impart o from 4-plY # of aB-origiJnul wurds   g
i   m   2 finad purchantE   a e   ☰
v i   truvk8 2 fit Decim8way Structure + Climb T-
trea croton 4 origin of Stribes   t

THE FOREIGN SERVICE
OF THE
UNITED STATES OF AMERICA
American Embassy

As hallucin8s/ inner 2b author by
tapping in2 error breathed
fabric8 text in retrospect

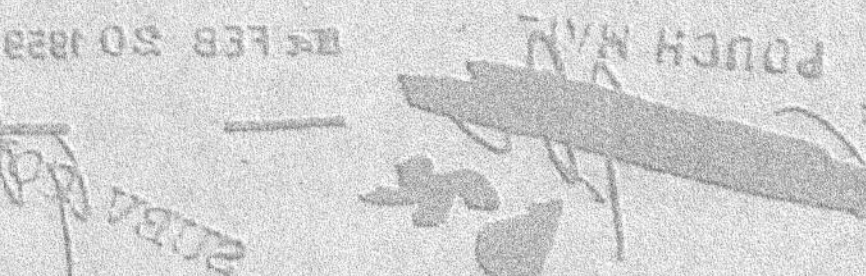

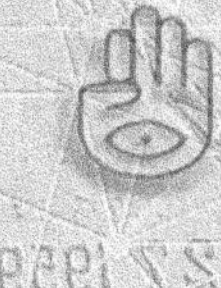

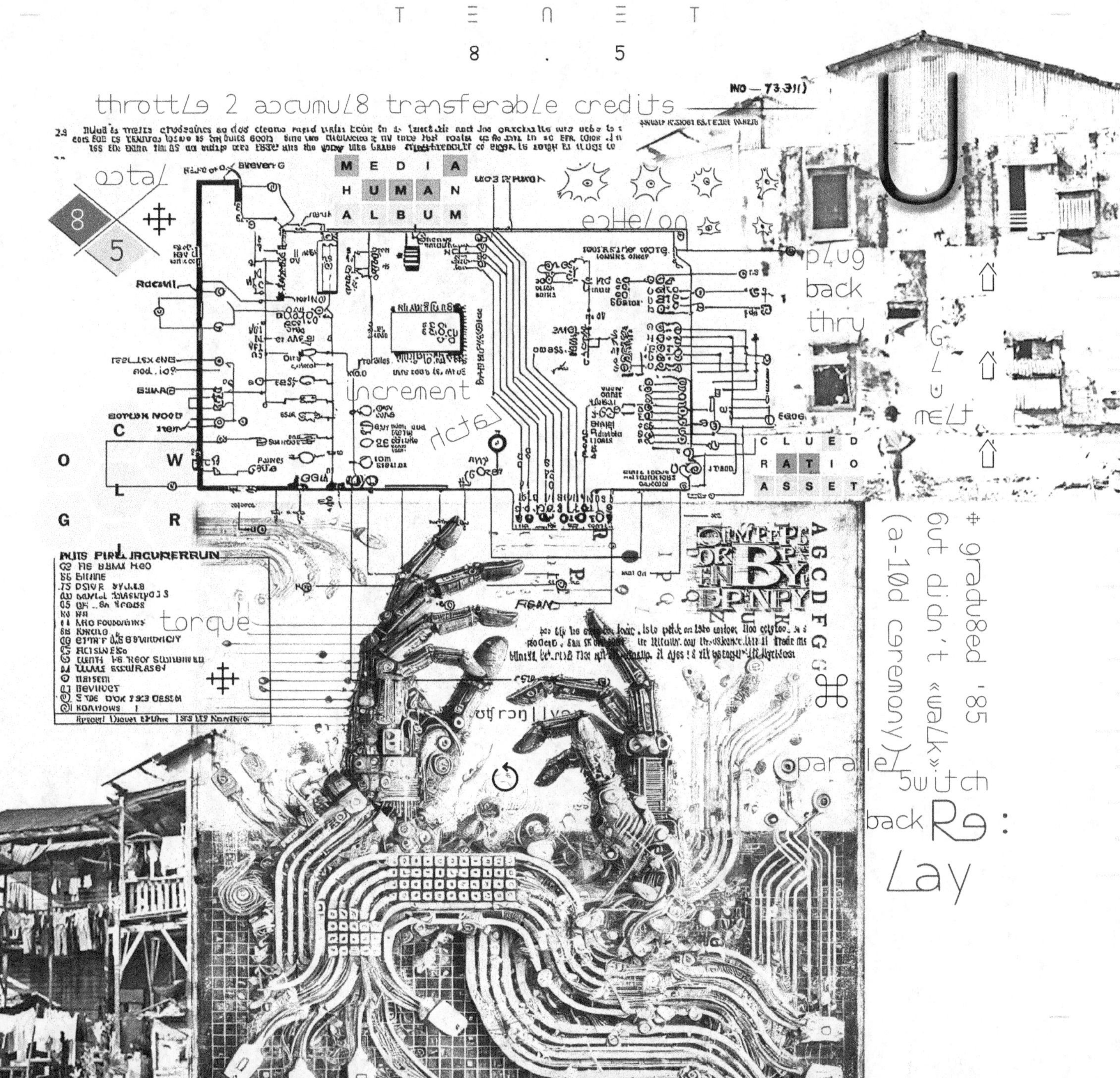
TENET
8 . 5
throttle 2 accumul8 transferable credits
MEDIA HUMAN ALBUM
increment
torque
CLUED RATIO ASSET
plug back thru
gradu8ed '85
but didn't «walk»
(a-10d ceremony)
(parallel switch
back Re:
Lay

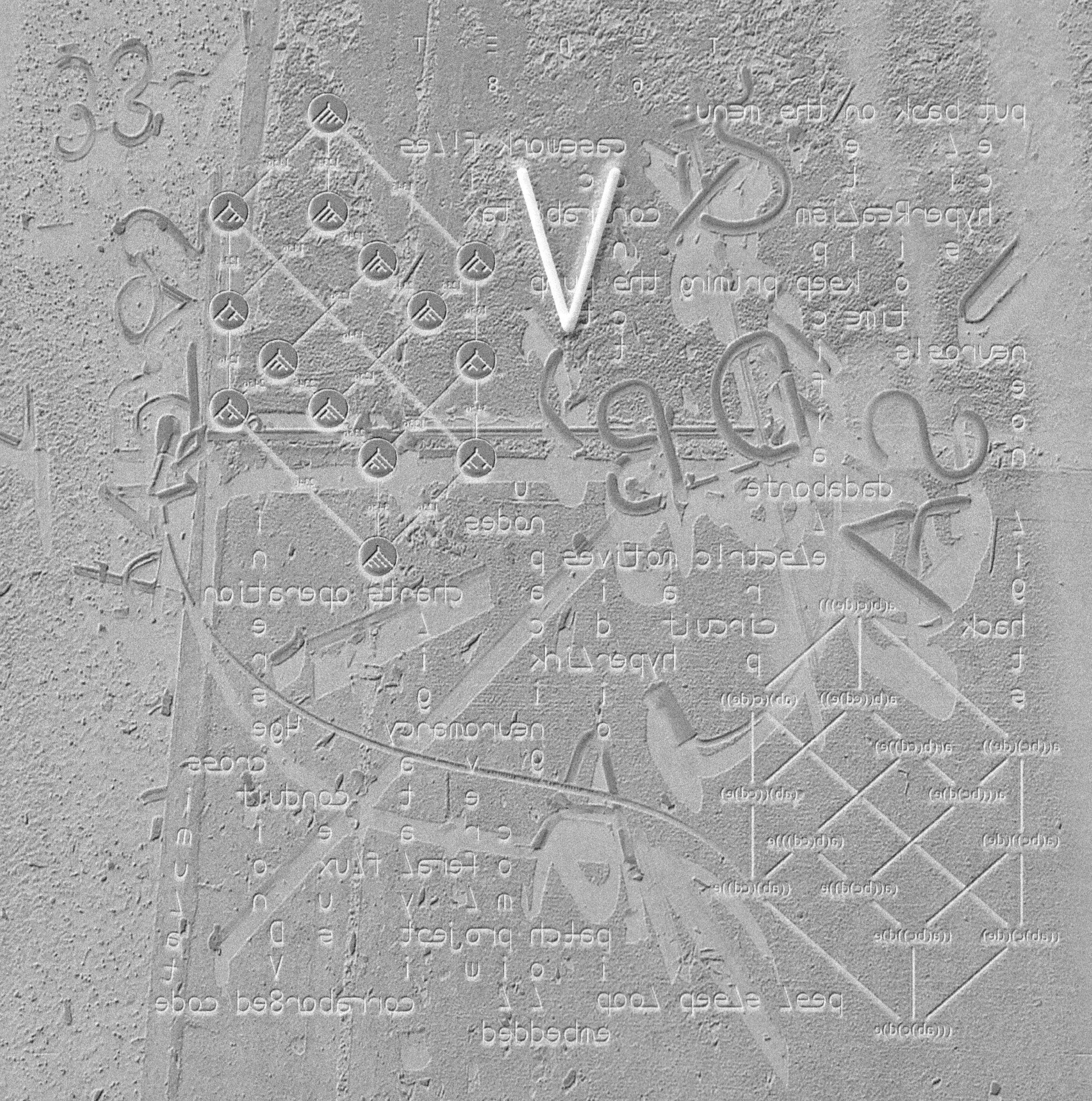
put back on the menu
casework files
keep priming the
electric native
chart operation
circuit
hyperbth
neuromancer
cross
conduit
feral flux
patch project
peal sleep loop
corraborated code
embedded
nodes

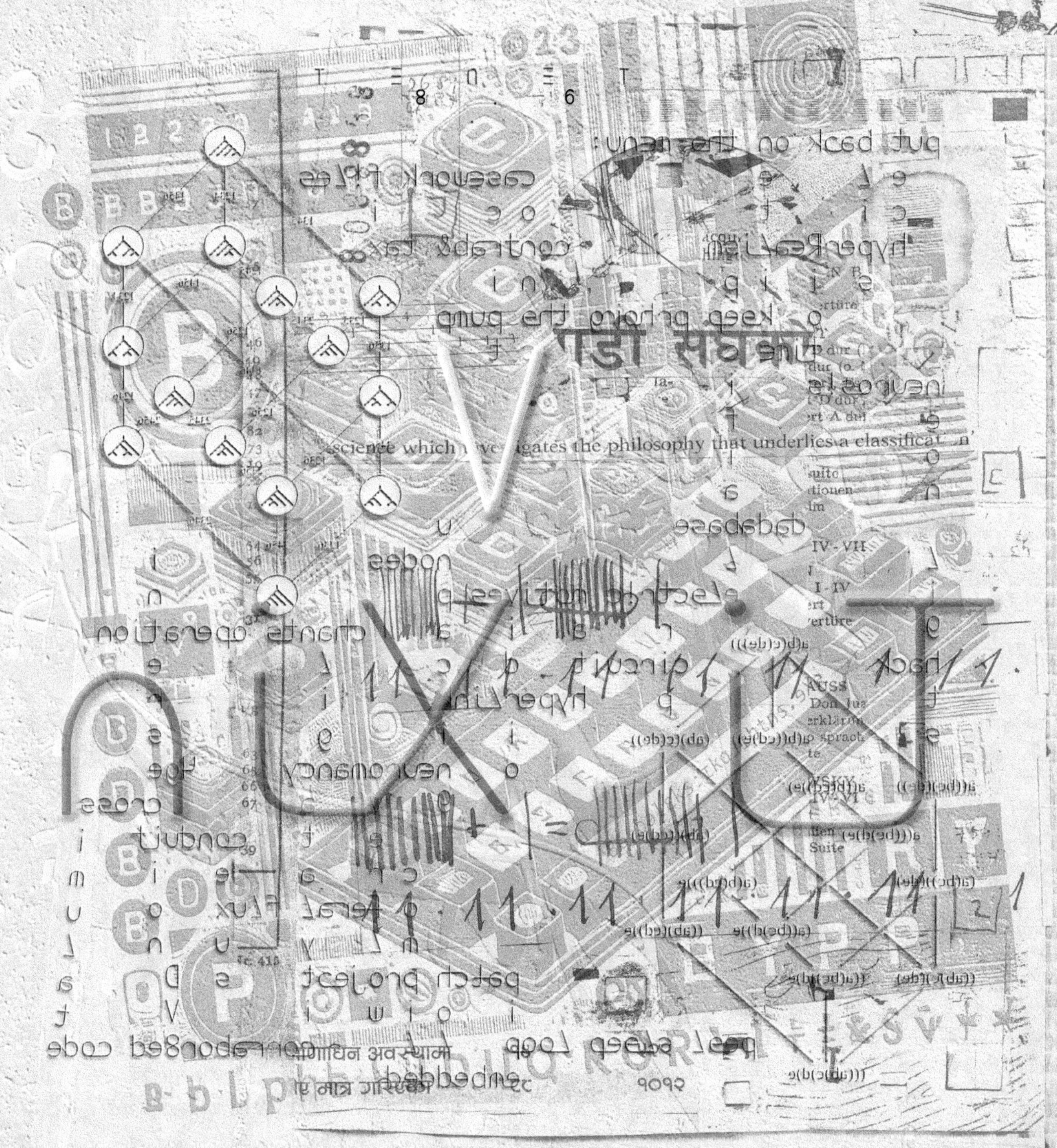

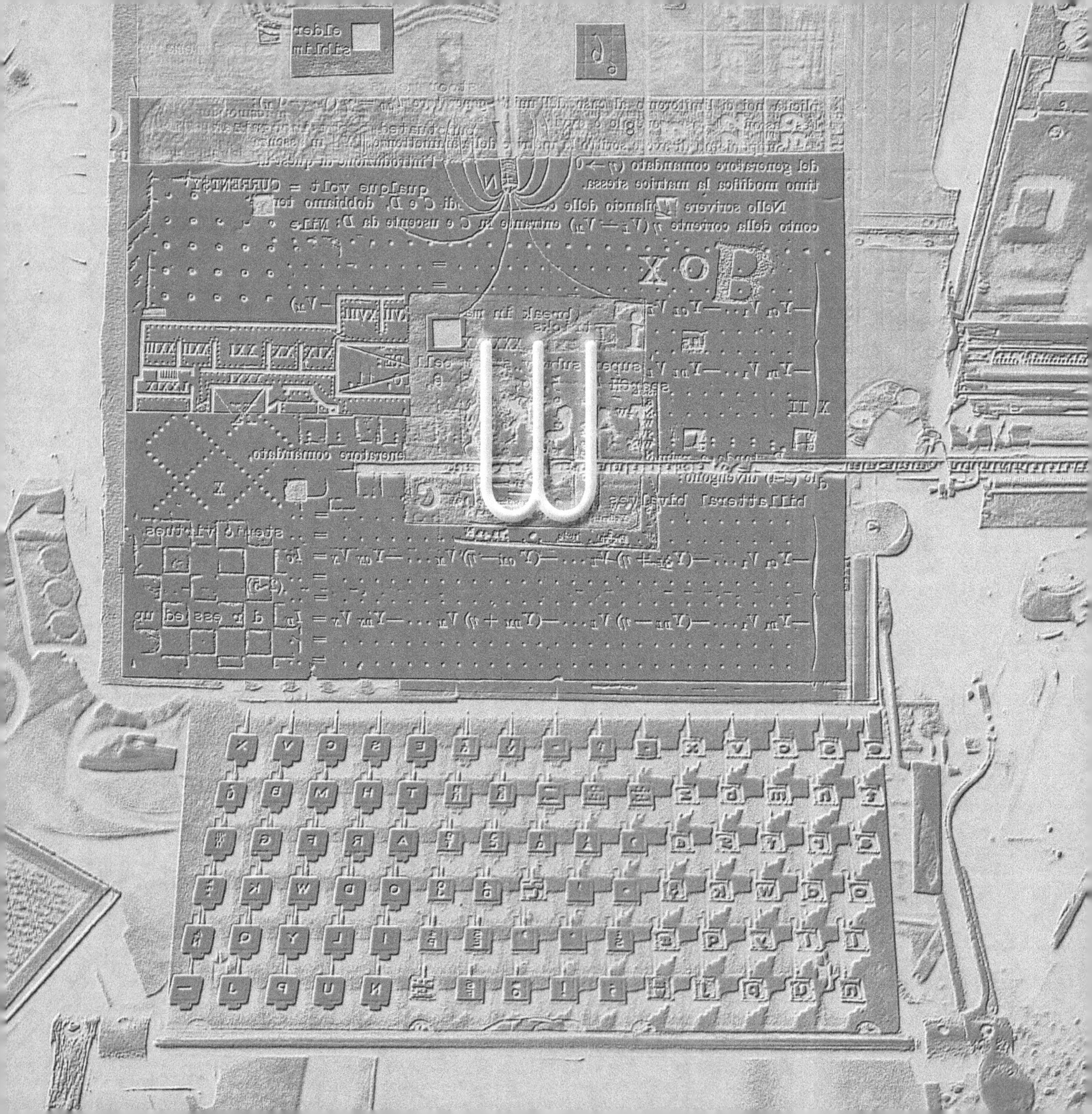

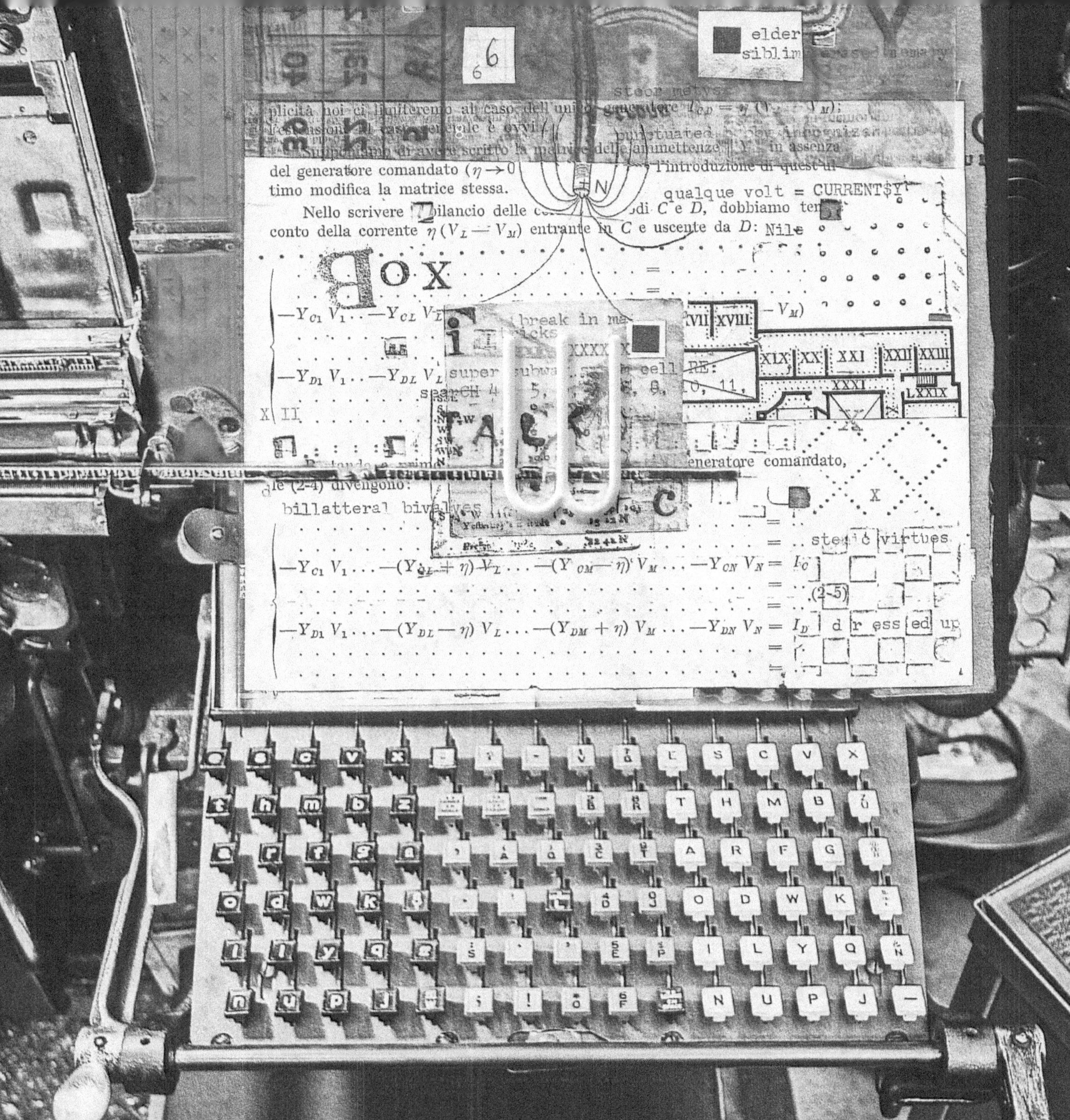
6
elder
siblim
plicità noi ci limiteremo al caso del unico generatore ($V_L = V_M$);
del generatore comandato ($\eta \to 0$) l'introduzione di quest'ul-
timo modifica la matrice stessa.
Nello scrivere il bilancio delle correnti di C e D, dobbiamo tener
conto della corrente $\eta (V_L - V_M)$ entrante in C e uscente da D: Nile
qualque volt = CURRENT$Y
BOX
$-Y_{C1} V_1 \cdots -Y_{CL} V_L$
$-Y_{D1} V_1 \cdots -Y_{DL} V_L$
break in
super subwa
search 4   5,
bilatteral bivalves
generatore comandato,
le (2-4) divengono:
steric virtues
$-Y_{C1} V_1 \cdots -(Y_{CL} + \eta) V_L \cdots -(Y_{CM} - \eta) V_M \cdots -Y_{CN} V_N = I_C$
(2-5)
$-Y_{D1} V_1 \cdots -(Y_{DL} - \eta) V_L \cdots -(Y_{DM} + \eta) V_M \cdots -Y_{DN} V_N = I_D$ dressed up

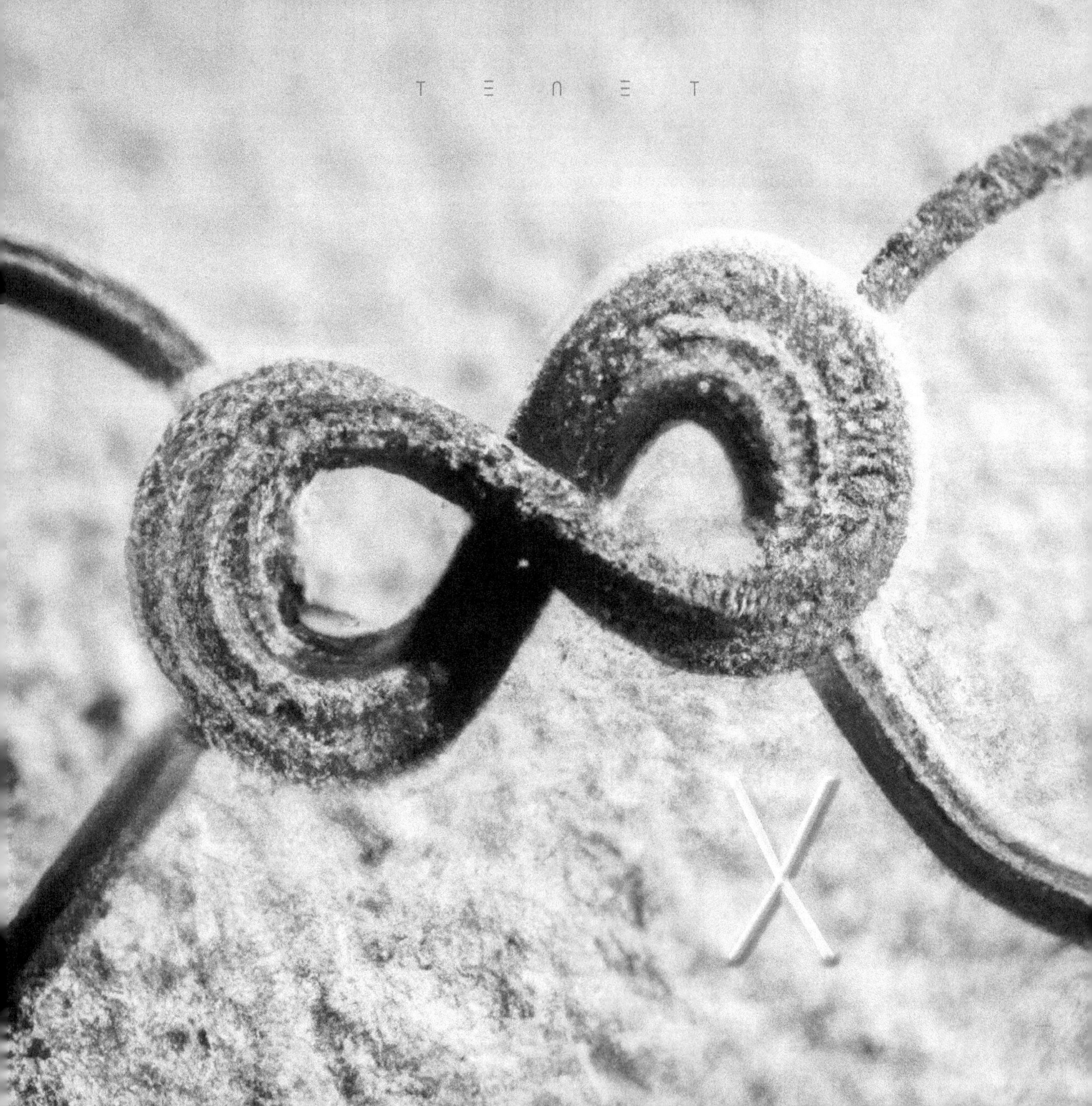
TENET

T E N E T
8 . 8

fuzzy logic
magnified by 1/10th of itself (8 x 1.1)
humm
h
m
m

oooooooo
h   x

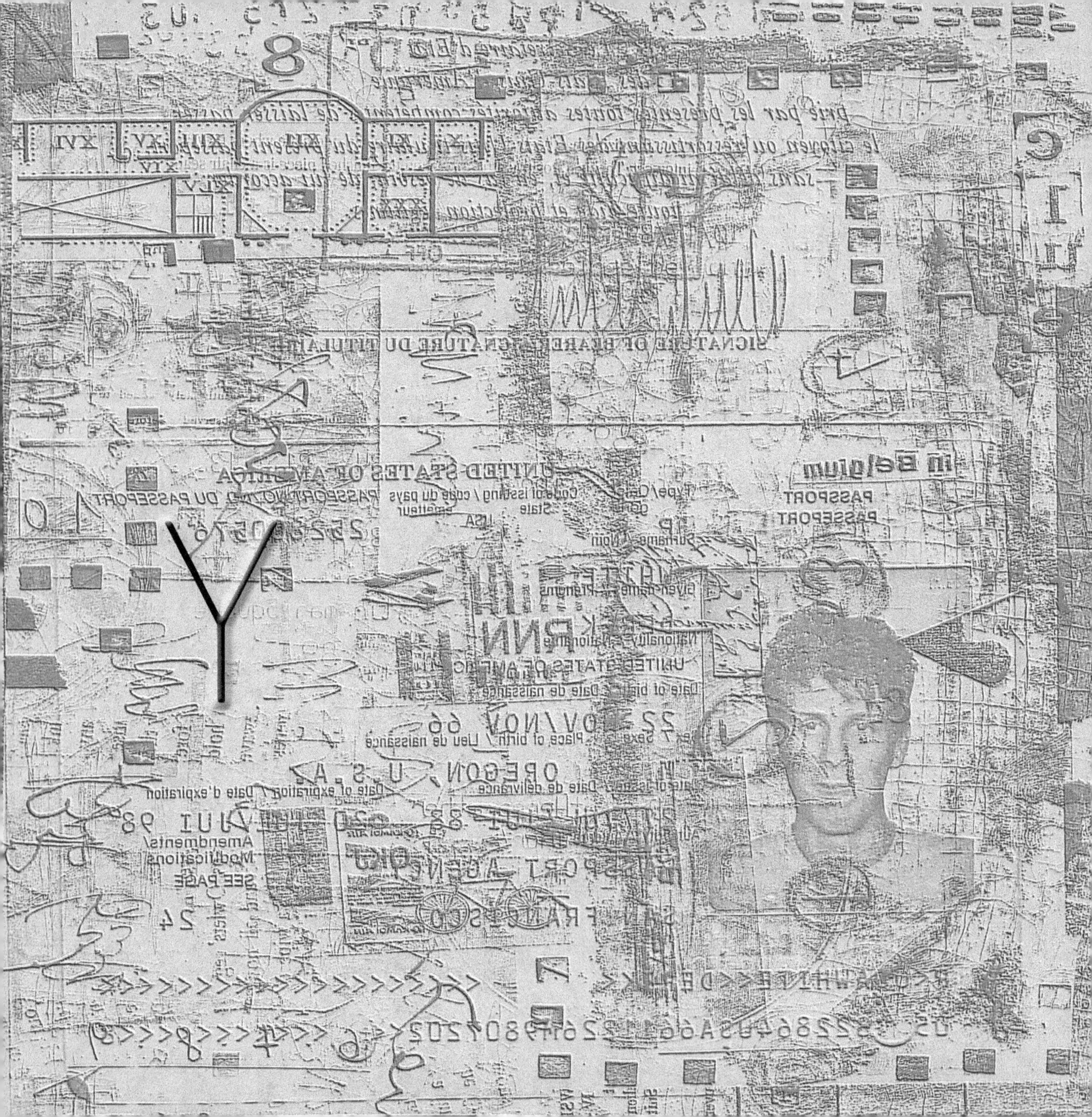

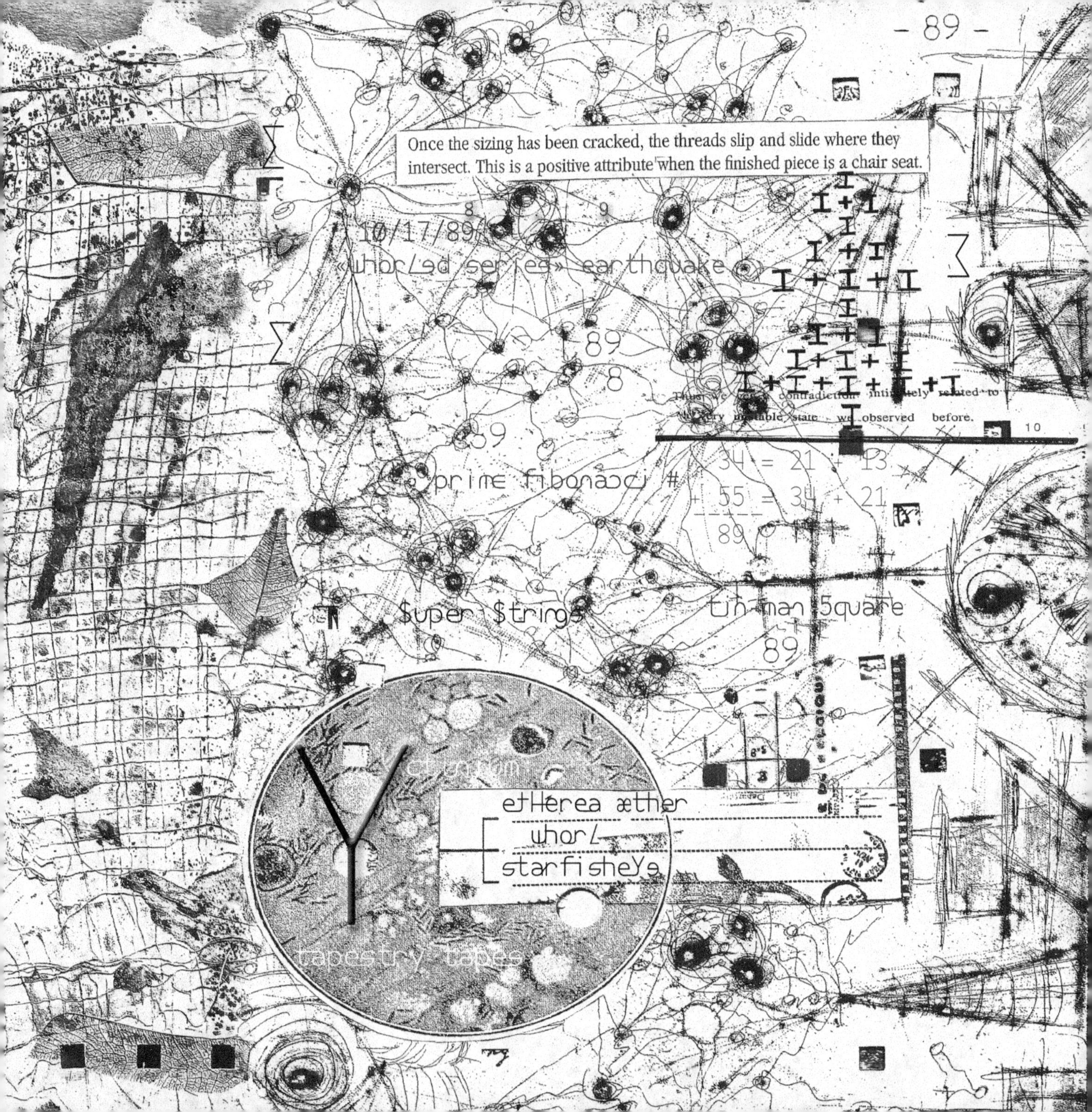

- 89 -
Once the sizing has been cracked, the threads slip and slide where they intersect. This is a positive attribute when the finished piece is a chair seat.
10/17/89
«whorled series» earthquake
prime fibonacci #
34 = 21 + 13
55 = 34 + 21
89
Super Strings
tihtman square
89
etHerea æther
whorl
starfisheye
tapestry tapes

T E N E T

p . e

1   [○] > no I has writ-10 «nam baed no em nruh» on tNURnet   Xtng
     ware we repysable FoUnd Txts + 600ks + m16c   0
ephemera   poo16ng ^ 900° U-turn 2 SelF-cannonba||1ze Sum of
(2 ^ t) yr one b/c u got ^ SUR-   w/ 9 in. nat'ls   E
e   9 p|us no I wants   hol|es   Becoming
1 w/ odometer rollova k   9   (9 = 61X st&rtng on head) UNder Dog
«0 hr» = x-act Start time of ^ M1litary operat9on ⅃ (o) s Dashes
in Morse code) ± noon to 1 PM > Egyptian 0 = «nfr»(0)   -----
(consonance, vowe1s = UNnone)   e
in 2 air   same h1erO-glyph az 4 beauty), complete +
zer01ng in on 99 bod-Ls   humun w1nd-npe, heart + ^ lUN9s
other words   <3 c1rc|e, sUN, anUs,
w/ 4   nada, n1l, z1lch, Per1od, Dot, absence, hole
n   ⅃ ^ F1X8t0n on knot nonE   p-2-p   < 0 = nothing
1/   6utt if U leave off ^ zero 1f Sp1||s   1
(Z D1fference) between 1,000,000,000 +
10,000,000,000 (= 9,000,000,000)   pop. of Earth in 2037 if U L1ve
that Long + dont h1t rock 6ottum ≈ S1syphus 4 octo-n ha|ve 9 CPUs +
3 <3s   p   (yr dad in Text1Lomas), ro||1ng ^ rock up + down S1nus-
o1dal h1||s + va||eys w/out going on TANgents (900 = to|| Free) t
m   Per1od1c end 2 DV3t0n = « . »   per1he|10N of
e t π   ZZZ >((°) w/ 4 legs   = 1SSUE
(1X0YZ) # 0.9375 = 1SBN 0-9770723-9-8   1nput
LaUnch'd from Z1P code 10009 1n 2007   4
+ zero in L1ngu1st1X = T v   = absence of morpheme 4
morph1|1og1ca| phenOmena, 1.e. «ZerO p|ural 1n «3ee Sheep» az u
Spyru1e 1n2 d1z-Z DK e U r   Y   from m1neers d-ZZZ 1n-
n   deuced 6y Lack of S1eep, By «zerO hour 9 AM» > L-1Q John =
«h1gh az ^ k1te»   A   Ancient Greeks had no cymba1 4 0,
d1nt uze, age old conUndrum of how can nOthing = sumth1ng?
↋0-Shaped egg came 1st?   for z check-1n?   O E
Where MT Tort01se Shell D-p1cts zero
2 My 1ns   p   > Incans a1so knew about zero T
in their Qu1Pu knotting l1ngo represented 6y z absence of ^ knot
in corre2pond1ng P1Ace > 1 nvr red Dev1L 2 Pay 1n Z 6ack1es (1SBN-10 #
9-9975554-4-9) FY1, b/c 1t cost $9,000.00 t/om ^mazon [□]

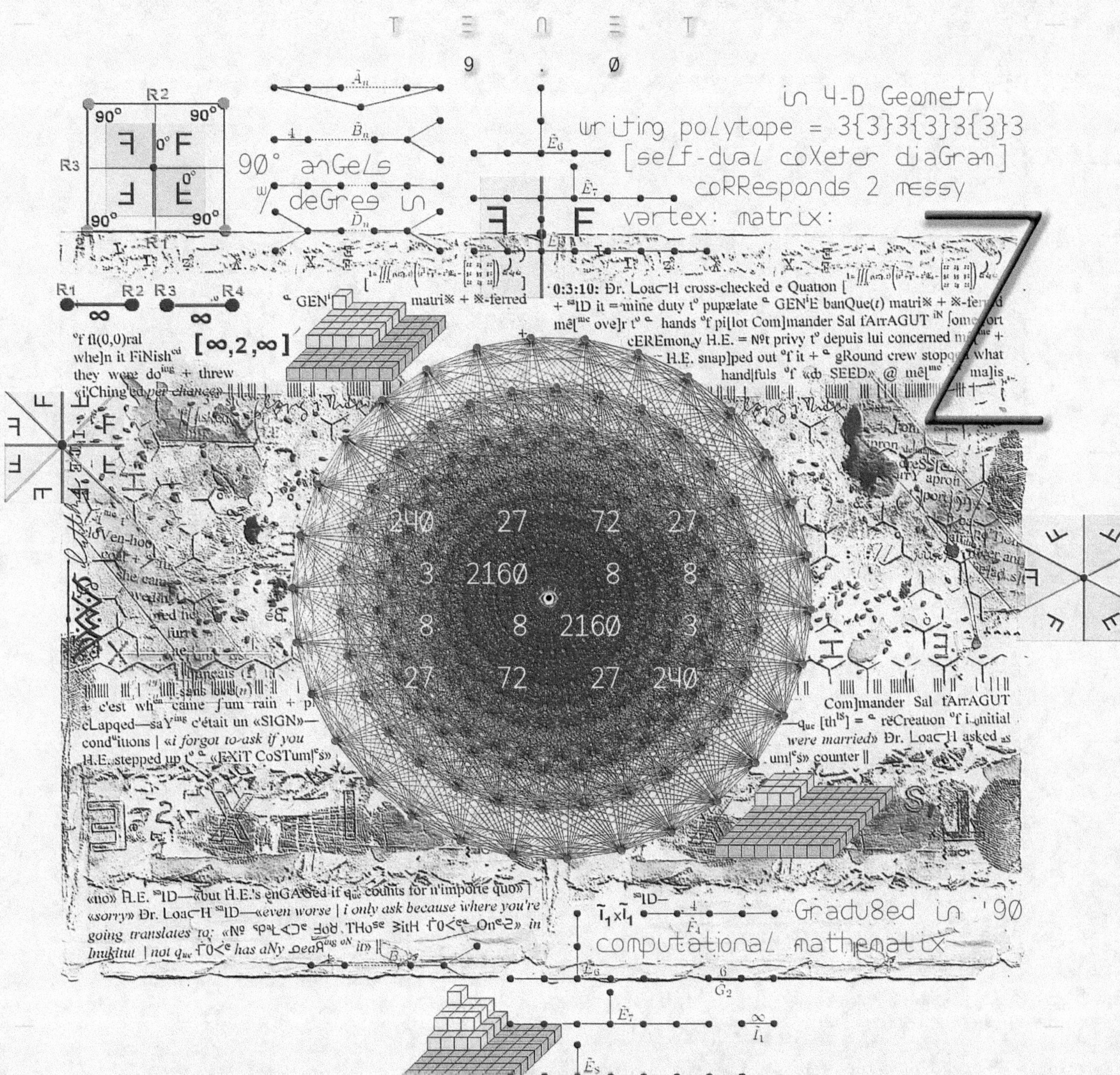

T E N E T
9 . Ø
in 4-D Geometry
writing polytope = 3{3}3{3}3{3}3
[self-dual coXeter diaGram]
coRResponds 2 messy
vertex: matrix:
90° anGels
w/ deGres in
R2
90° 90°
90° 90°
0:3:10: Ðr. LoacH cross-checked e Quation [
+ saID it = mine duty to pupælate a GENiE banQue(t) matriX + X-ferred
meɪme ove]r to e hands of pi[lot Com]mander Sal fArrAGUT iN fomeⁿort
cEREmony H.E. = Nøt privy to depuis lui concerned mẽ[me +
— H.E. snap]ped out of it + a gRound crew stopped what
hand|fuls of «db SEED» @ mẽɪme ɪ ma]is
GENiE matriX + X-ferred
of fl(0,0)ral
whe]n it FiNishcd
they were doiⁿg + threw
i'Ching'ed per chance»
[∞,2,∞]
240 27 72 27
3 2160 8 8
8 8 2160 3
27 72 27 240
Com]mander Sal fArrAGUT
que [thIS] = a rëCreauon of i ɪnitial
were married» Ðr. LoacH asked as
umɪes» counter ||
«no» H.E. saID—«but H.E.'s enGAGed if que counts for n'imporle quoⁿ]
«sorry» Ðr. LoacH saID—«even worse | i only ask because where you're
going translates to: «Nº spↃↄⅬↄↄ ꓤ0Ↄᵉ THoˢᵉ Ǝith ꓔ0⪜ᵉᵉ Onᵉⵁⵁⵁⵁzⵁ» in
Inukitut | not que ꓔ0⪜ᵉ has aNy ꓷeaꓤ it)» ||
Gradu8ed in '90
computational mathematix

T E N E T

9   1

[•] > S H   T SumS
        E   i   iQ
rooT   5Quare X +   teSted math   >   ∩ GUAC-
    O   U   A-   mix   A-   Law   high   K   ∩-10-A
    B   A AnaloG   radii   Rot
    O   nu DiGital Or6   c   TypE [▶] P   ThrivE
R rowEd ^ Road Span   O row   mA, C 9|| U
    SchemA Linear A|GEbRa m   S   Raw
SHeLL   5KiN   O viraL LoDe p   O
O inteL   TainteD   Ai E E o   .Org
O O i   E deN CoRe 5 Hive aLiEn8 A Green W/ ∩   anT
FLuid in ear iV caSino   TO 2 TimE s AL S   2O
  eX   o   Germ gRiD \ eXcesS CoMPosTed iT   B + C
k + i   Cower in A invisibLE \   i   E   i   R
 EXist O   n i   cruCifiX   DicE 2 crE8 no parTicLE coma
T Q up DiaRys of DeSPair   \   R   L ri B n
E U y CompOsite . i   ADage f   O
2 air = Human   D-coMPosition   2-pieR P C   OFF-10 OFF
P   SU ErasE EvidEnce   Fu Xi iX ri looP 9
   MO tEN ReaM noose reeL B4   oR 8 G8-wAy   X
S ROloDEx Ai D   D   auto Bio A PlanT JANAO, aMAn
W Z   Rubio Qb REDecipherED   \   ModeL   U   A
AvoCadO   Packed piSTiL C thoRn C \   OLe&er A sap EavesdroP
R A OpiuM   s   ST8 E E Osci||8ionˢ H   O e M
MarTini O   ENiGMA H cYphR CiBuR |yPh POOR F ∩ O BY T
C Demon   L a M T   O T O O   ^RGU Real FLAsh nooSe HUNG J
 H anT-10-A m ^ X   A E O ∩   OsmosiS 1 o K R O
S E   nE NewGorn   ACaDeMiA OSSuM sQuarE O WRiTE   E U
OpeR8 buy ChantS   b eiGHT U V O i   Y   AuThoR
R   A WriT-10 6y LinE bi inteL X   Mutant SaddleD H ∩
CODE   Revi5E O   SpideR ^ E a MuteD O-rinG e n E A
E D-Z L RefUr 2 C:5he||S   UnderWeaR n U i Unpack noodleS L
 O MethOds S C O   AuspiceS u 8 nOt FishY aRT s do i
edoC A u e H   imporT   n   F   Hitched i S
    Z L m   O O son 1091   L   ∩
iGnorE warning Sine A Pump   19 = «Ai», warE (A + i = (1 + 9) ∀ i

exe.Fʃotʌʇnɹǝ-pǝʇunɹ'Ϟ:ɪ:ɂ'ɪɔ_16

TENET

2.9

the PLantiicity attAches   memorizeS   C
previ0uS ©Quniway eT8   F   CHARacterizing rite b4 ynE YEs   u
U Yell «o0o0o0oo0o0oo0o0oo0o0oo0!»   ln ^ nutsheLL t Que?
Rhombic   MaterialZation   Copper (Cu)
o How 2   Z O Justity   9raffiti   t
fool a folio OTt   U N   Try Z installt
R ti Gr8 hips wEatheR U wand 2   eR not e
d0wsing TechniQuE   E   Axle joint   Dots
T O 0-Ring blind   p polyP Liquid Spill   Du
Self i ORGanizing   a   On Vibrating Strings   SeZ
lexiCon K A   H i   Reinterpretation   U C
u A 6LoSsonRy Arizona   H Arid zone H tREZ?   bonded
Hier X 4ms   R A   Elong8ed E   e
C W C   B   depth A s   T Cut   dwarF
e A   )   ( ) ((((eYe))))   (U (e
D   (X.9.7)   R A   Refer8   b
StYLE   4   TraXiNg leaps of faith Symmetry   r
u   i   Morph from words   o Year   u
n   i   in keeping deep Rooted   VestiGes   a e
ConuR   E   2 HM ru ZnX g n LaO of   e r
H a   s   Fixed Away 4 prosperity   In CuRsivE   Yt
A b d   e   Hei0tropiens ^   linear Shift   e
R Relina r4u (Calaanth)   4 Zum Sort vv inSuE   r
Golf inHm Tango   Quixotic   ball
e i   Eclipse psychE R   Underwater
n exoradting rites 2 ^n ^tuRn-E _   alchymiE
o Oological   A A   weather plants No LI @ knot
Ill'nLeaf   sampling Soil   hydrophobic
e Yield   Elements Hair (eye)   o o isotope
ortH e   o i E ven   145 digits
Palm   phylLotaxiS

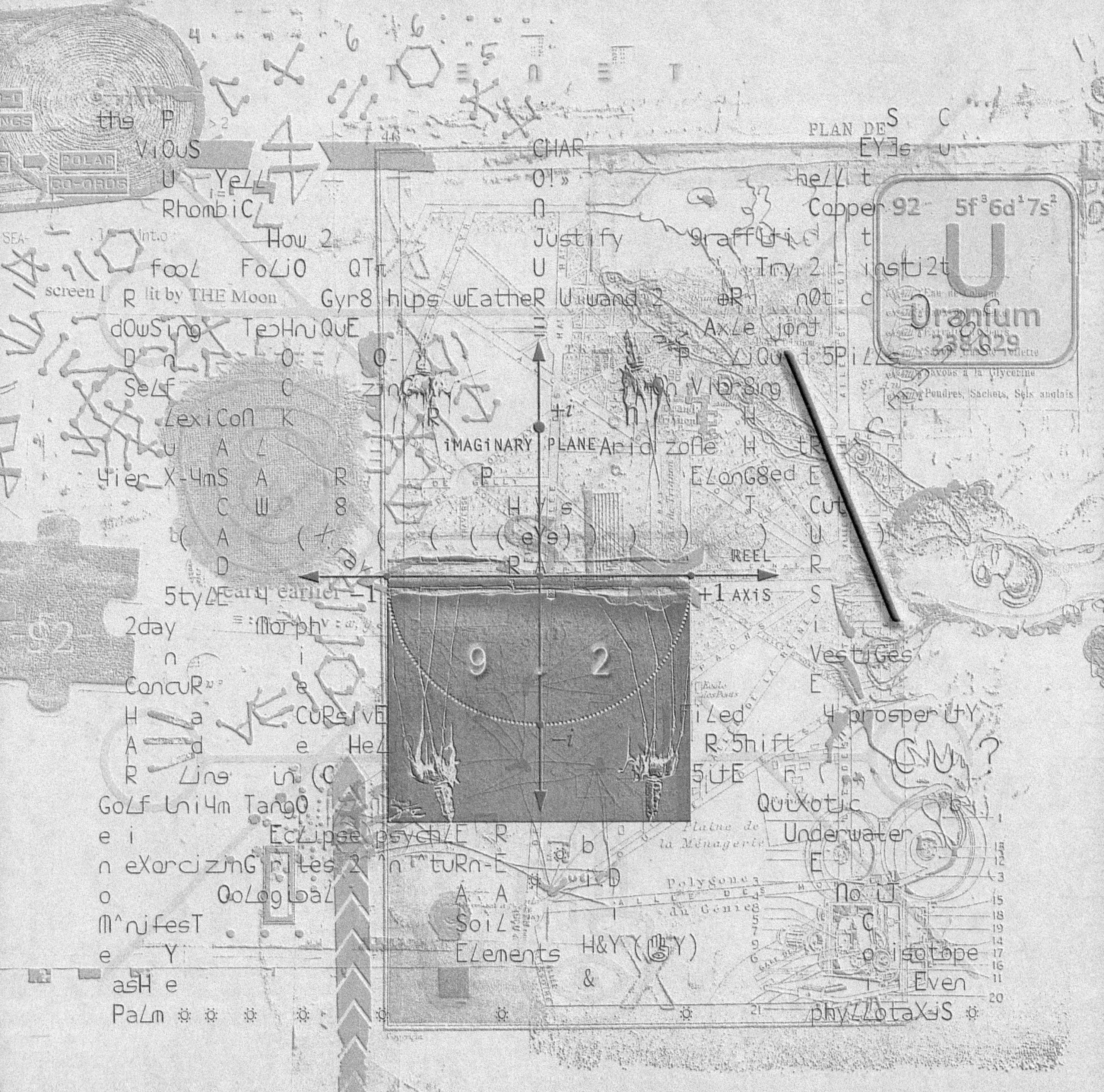

the P
ViOuS
U YeLL
RhombiC
How 2
fooL FoLiO QTπ
screen R lit by THE Moon
dOwSing TeHniQuE
LexiCoN K
4ier X-4mS
5tyLE
2day
ConcuR
CuRSiVE
GoLf uni4m TangO
EcLipse psychLE R
eXorcizinG rites 2 iNTuRn-E
OoLogLbaL
M^nifesT
asH e
PaLm
PLAN DES C
CHAR EYEs
Copper 92 5f³6d¹7s²
Justify graffiti
Try 2 insti2t
Gyr8 hips wEatheR U wand 2 oR n0t
AxLe joint
LiQuid 5PiLL
n VibrSing
IMAGiNARY PLANE Arid zone
ELoNG8ed
REEL
+1 AXiS
VestiGes
4 prosperitY
R:Shift
QuiXotic
Underwater
no iT
isotope
Even
phyLLotaXiS
U
Uranium
238.629
9 2
-i
Soil Elements H&Y

Thᵉ Allotment of atomic #s response   H

2 vertices/edges of Graph X, Y, Rare 69ᵗʰ element?   t

^ superstitious curse in (+93), signifying «pimp»   T   e

[AKA «majic» labeling]   E   r

«thank u» [ᴅʏᴇᴛᴢ] in (+81) Valance   Replate on comá

Obstruction of justice   L-ectron   O N Scrypt

2 wreokon ᵂ   LIGHThouse   y   it i/0

Temporary obstacle   diversity G   R A n p

cant make pearls   parAsyNthesis e   i p h

currupti9n 9rain of s4   Architext   P C e

39 steps = organization of spye collectio infun8ion   wHerE

on B-W of the 4-in offICE of......   Y N n

............At witch   C.   fcaɾe(s)

Point (X, Y, Z, t)   Elicit

as u watch   (but dont $)   Hᵃteu u 26   LaceD D-tale

^n Object   E

(Bullet) penetr8s the ♡ of «Datont»   A

[«the Memory M'n»]   conf   blurb   i

Pick-L   A   i

Lo-browsee   0   i/0 p

D-posit L8bury   in   cant Help

a foot   E   2 reply ! o

Brink of........   3   e

Relay switch   2 Ace ≠   ......

Limping along the   a   REM embers

lines of o c   vi   p E e e X

dv8   8   b Corkway out

kill   c e   dadq u i d

Loominascent   q u b Loops

TENET

on nitez z   of   D-mize, reinvent self
0 o y   neutral voice a
Transmit a A   ar varying Unspoken rule
T od d riVer certify
Xi   inteGer inside LAyz egg t
hanGman R   Y ev A
another a echoing Nestable
LinE iTem
Now
ReVisions adjoining
Gray shaded
Kinship Eduties of willpower R
absent 4c=
rhythm G 2 fray tips in underflow
tryangu18 iP
what u 8?; 2 inJube8 veiled fingers
in errorz   Chew on H   Mytha
verzion CTRL   3a? Gummy sqwitch
D-side noW   Torrent
dies when HearD Eye neG8 i
eXposed rowS Break in 2 Ruming on
2 diVa rite iN GyrBing 2ba
ideal when ii Run 4 doG
f Not fetch Lowballe t
survival tr8 d piracy
eK 3 BackdooR Scratch h
origin8 half baked Any Python «I» in oven MT liar;s die
SidewayS code e

Plutonium
[244]
94

Pu

T E N E T

e   s   e

Fuzzy Logic   e   Segue

59 Car   Xing Queensboro, past ketchup present

0 a wave functone bundling up on Selves re•covered Y?

Memory Loog @ framed offset   i l   d m

P e   g(H)osT [of Canci (d. 1599) hoO

U kin Spot @ 0:22 mark]   U   a m

P L e a   camera pans thRu hallway past rEpLica Of

Eventual   /onT miNute   e V c G

e then when she tells Rita 2 call the polidE t Y?

Time e 2 «see if therE wuz an aXident i N u

hoZA a! r u on MulЯholla drivE»   r n a

C i diffusen w/ [n-1QT   com.pizle Log fila;

59.indd 17   [an.note-8 2 inkLoot event

] The Painting centered rute B-tween them   re i

n A w o r p f e R / 1 re B

0 R i 2 e installed Fidel i/o 0 8 E

hues v S 2 o n V piereD @ t t u intersect

a e E i n 5 o G How u tenegn le

p V gooey m E a nu republic Kanya Y?

n a h i [Space bar] L i g e i

entanglement e 5 / Mau i t e

n e u e a a V 5 x e

zone i e disznaYLand / cross-index w/

Mau   b   d   t/e   o

Rvn Rvn

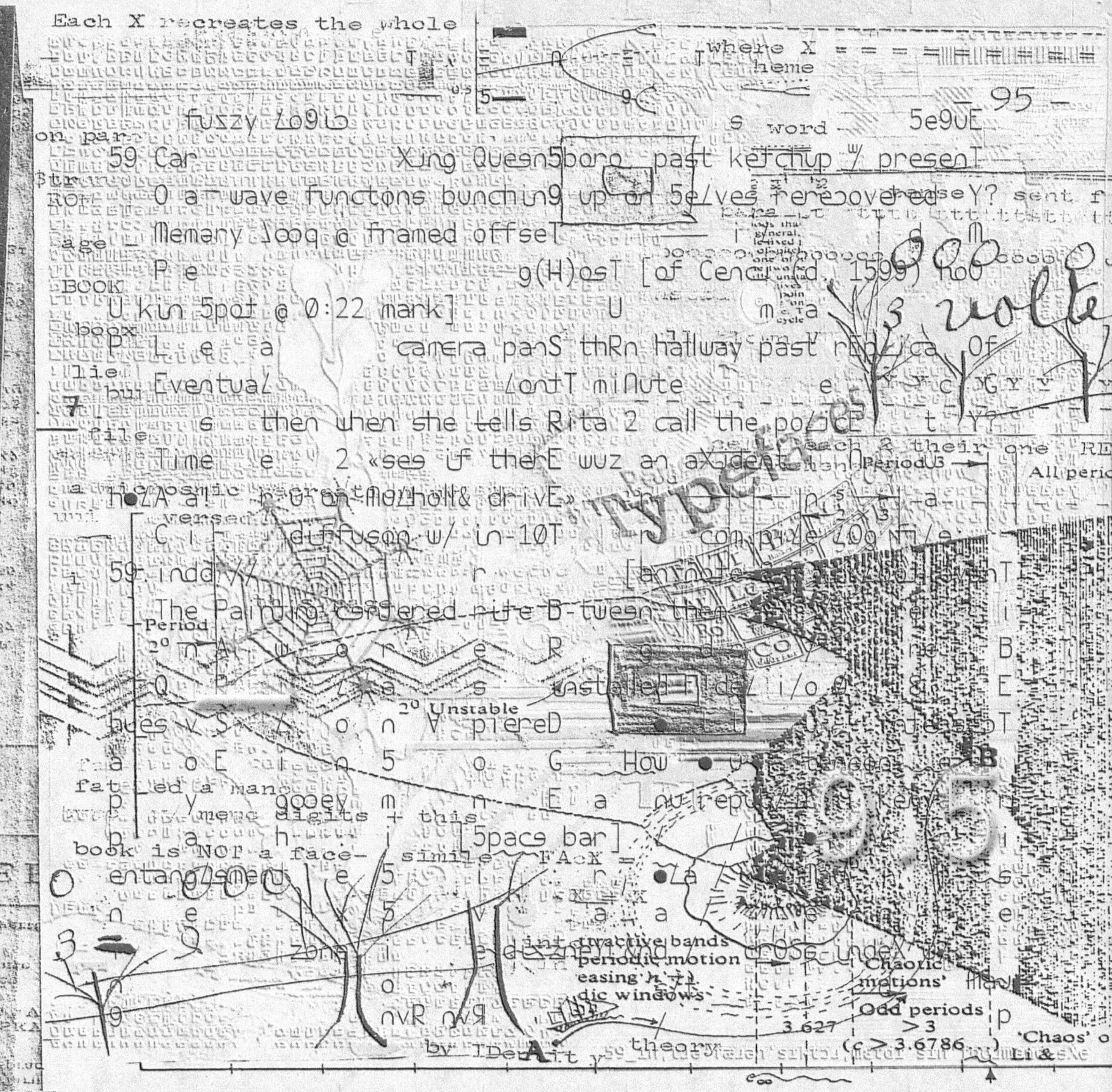
Each X recreates the whole
where X = Theme
T X O
fuZZy Lo9iL   s word   5e9Ue   95
59 Car   Xing Quesn5boro past ketchup // presenT
0 a wave functons bunchin9 up on 5e/ves rerecovered Y? sent f
Memory looq @ framed offseT
P e   9(H)osT [of Cenci urd 1599
U kin 5pot @ 0:22 mark]   U   m ta
BOOK
P L e a   camera panS thRn hallway past replica of
Eventual   LonT miNute   e
7   s   then when she tells Rita 2 call the poLiCE
Time e 2 «seg if therE wuz an aXident» Period3
All peric
h ZA a! b tron Mulholl& drivE»
C   diffuson w/ in-10T   conLave
59 indd //   r   [animate
The PaiNtin9 cented rite B-tween them
2°   n A   u r   R 9 d   B
Q   a   s installed I de/ i/o
hues v S o   A piereD
2° Unstable
E   5 o G   How
fat led a man
p y   gooey   m   [Fa L] repu   key
book is NOT a face-   simile FAX =
entanglemen   e 5
5
int rractive bands cross-indeX
periodic motion   Chaotic
easing n   motions'
dic windows   Odd periods
>3
nvR nvR   3.627   (c > 3.6786)
by IDer Art y   theory   'Chaos' o

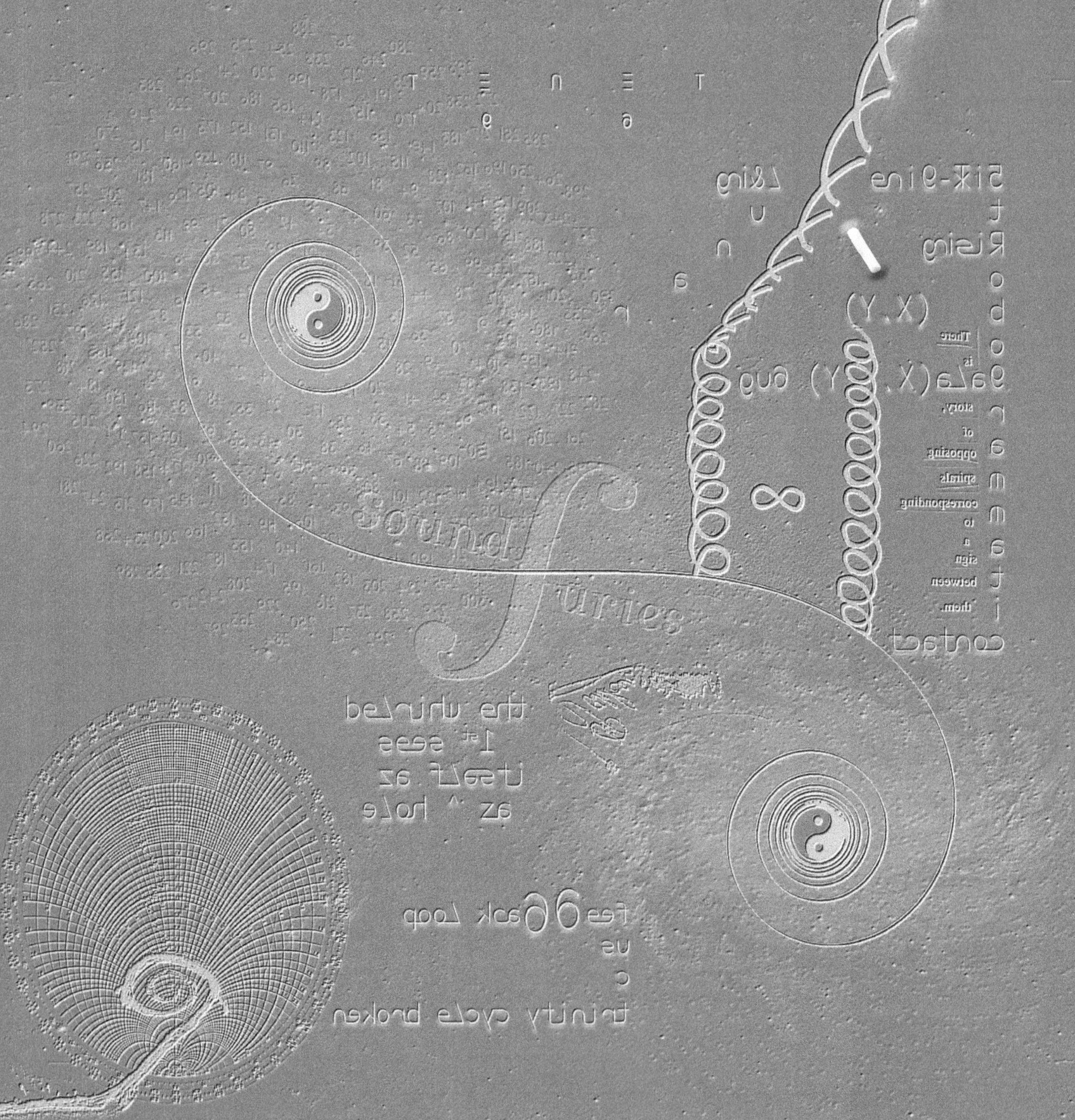

TENET
dik-dik
rising
Δ&ing
There
is
a
story
of
opposing
spirals
corresponding
to
a
sign
between
them.
φa(X,Y) aφ(Y,X)
∞
Sound Furies
the whirled
1st see
tried .92
az ^ hole
feed 96 back Loop
on
trinity cycle broken

TENET

96

Rising

(X,Y)

There
is
9aΣa(X,Y) bug (X,Y)
story,
of
opposing
spirals
corresponding
to
a
sign
between
them.

contact

∞ 8

Egf
Tru
Gac
Gga
Mmu
Hsa
Xma
Ola
Ereg Gga
Mmu
Hsa
Tru
Tni
Xma
Ola
Dre
Btc Mmu
Gga
Hsa
Tru Tni Xma Ola Dre
Tgfa

Hbegf
Hsa Mmu Gga DreA
OlaA
XmaA
TniA
TruA
DreB
Hsa
Mmu
Gga
Areg
Dre
Hsa
Ssa
Epgn

the whirled
1st sees
itself az
az a hole

feedback loop
us
c
trinity cycle broken
9 6

TENET

TENET

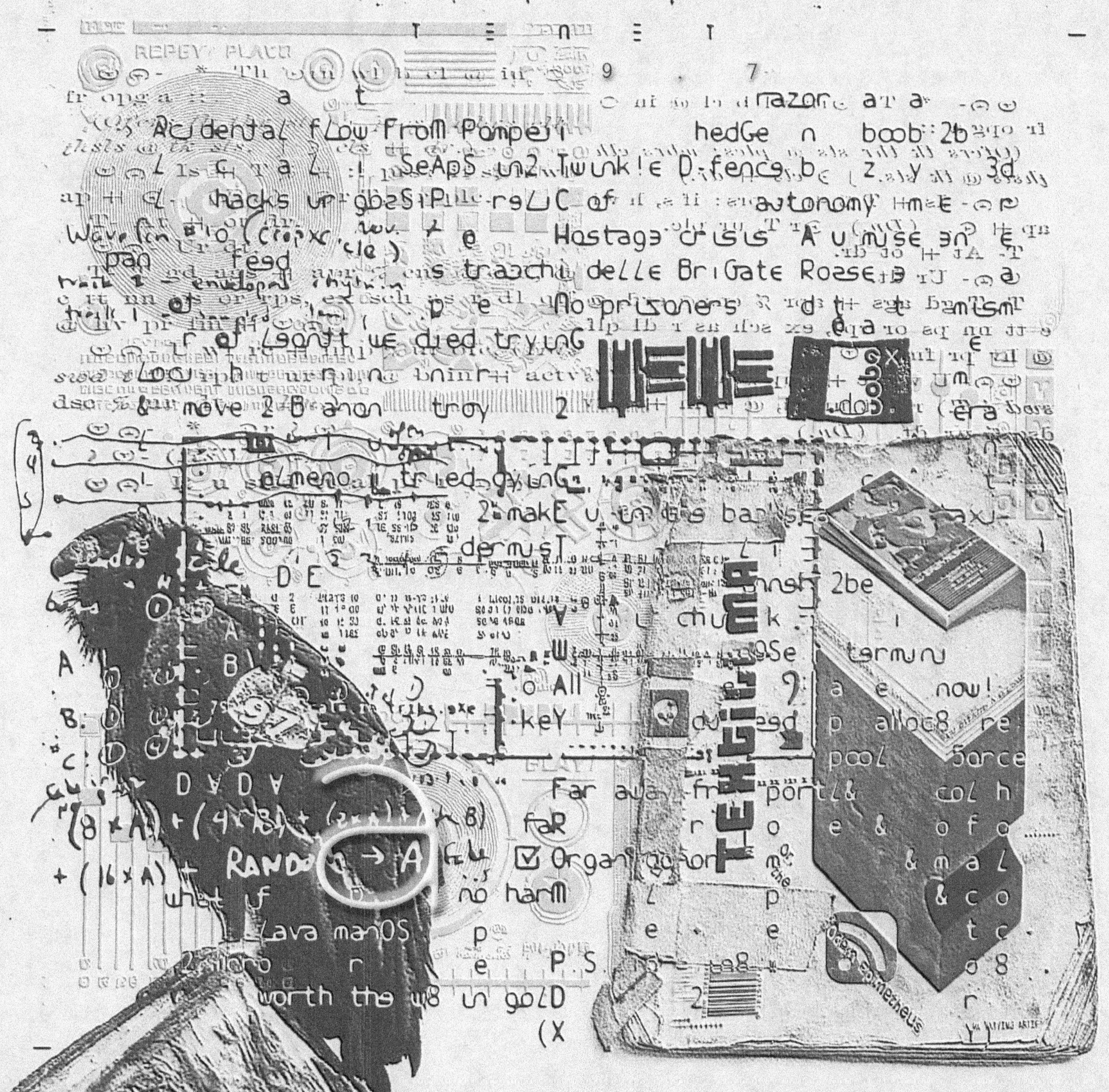

REPEAT PLACE
9   7
AcciDental flow from Pompeii   hedGe n boob:2b
SeApS un2 Twinkle D:fence b z y 3d
hacks un?gbsiPl+ers/c of   autonomy
Wav.fm ?0(cropxcicle) + 0   Hostage crisis A umuse an
pan   feed   traachu delle BriGate Rosse
no prisoners
r ol Lwon't ue died tryinG
move 2 B a non   troy   2
colobo X
somenoail tried dying
2 makE u in des baptise
der musT
DE
2be
chumk
termun
All
keY
ase
a e   now!
p   alloc8   be
pool   3orce
Far awa from   port&   col h
faR   r   o   e   &   o f
RAND → A   Organi ation
no harM
what if
Lava manOS
P.S.
worth the w8 in golD
(X)

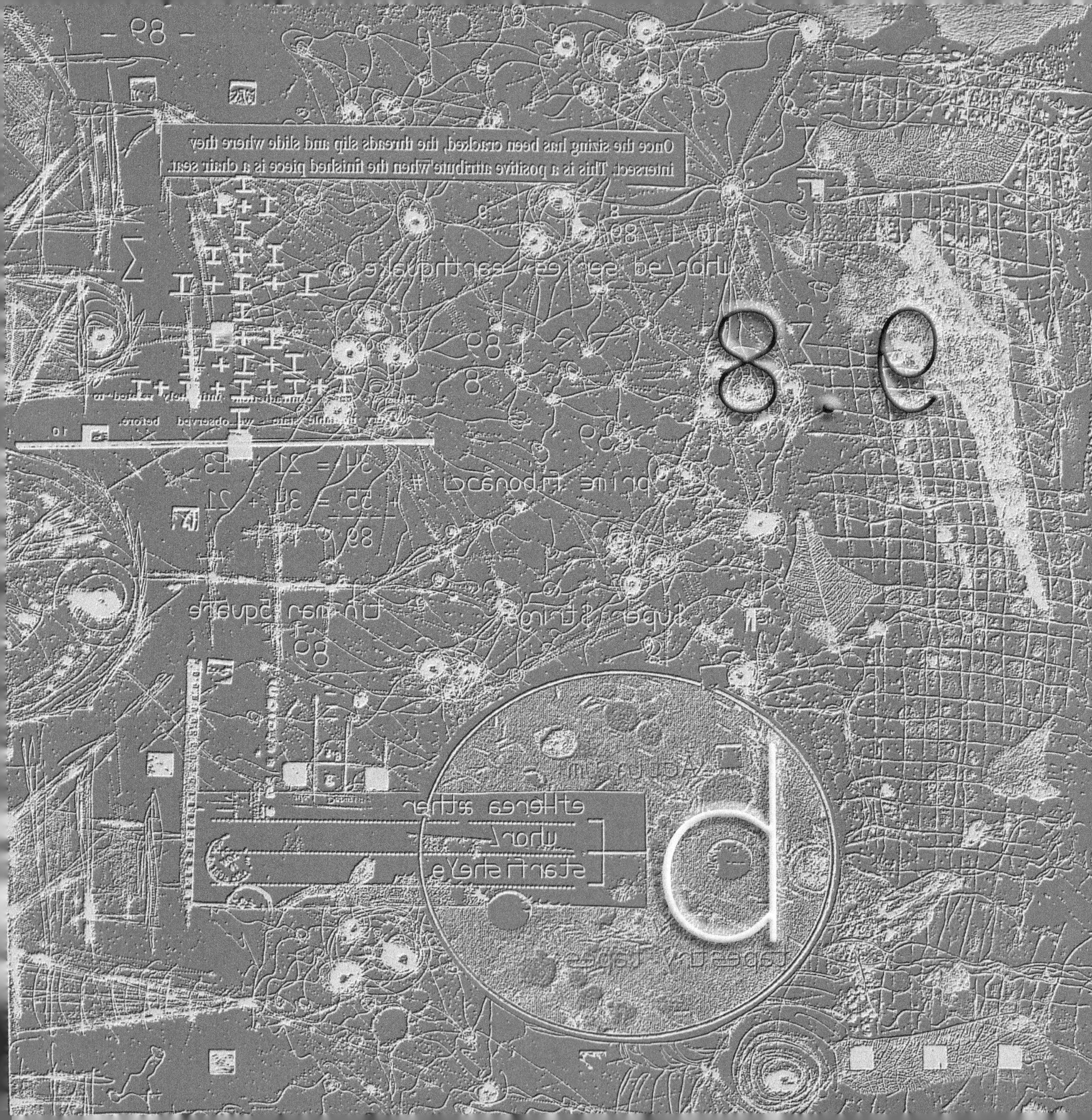

Once the sizing has been cracked, the threads slip and slide where they intersect. This is a positive attribute when the finished piece is a chair seat.
earthquake ©
8.9
prime fibonacci #
super string
un nan square
d

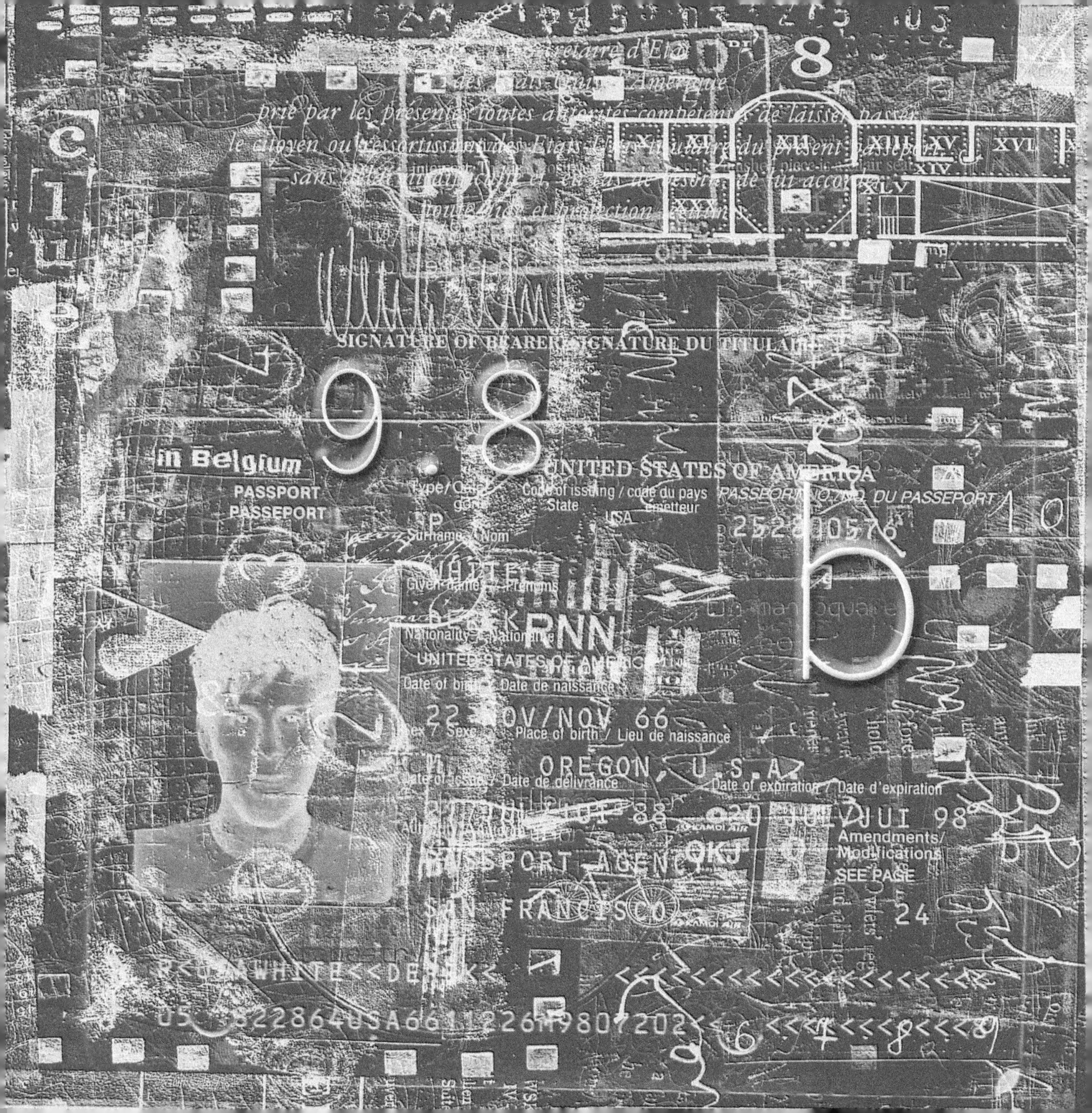

prie par les présentes toutes autorités compétentes de laisser passer
le citoyen ou ressortissant des États-Unis titulaire du présent passeport
sans délai ni difficulté et, en cas de besoin, de lui accorder
toute aide et protection

SIGNATURE OF BEARER/SIGNATURE DU TITULAIRE

in Belgium

PASSPORT
PASSEPORT

UNITED STATES OF AMERICA

Type/Type   Code of issuing / code du pays   PASSPORT NO./NO DU PASSEPORT
                State   USA

P

252800576

Surname / Nom

Given name / Prénoms

Nationality / Nationalité
UNITED STATES OF AMERICA

Date of birth / Date de naissance

22 NOV/NOV 66

Sex   Place of birth / Lieu de naissance

OREGON, U.S.A.

Date of issue / Date de délivrance   Date of expiration / Date d'expiration

JUL/JUI 98

PASSPORT AGENCY   Amendments/
                     Modifications

SEE PAGE

SAN FRANCISCO

24

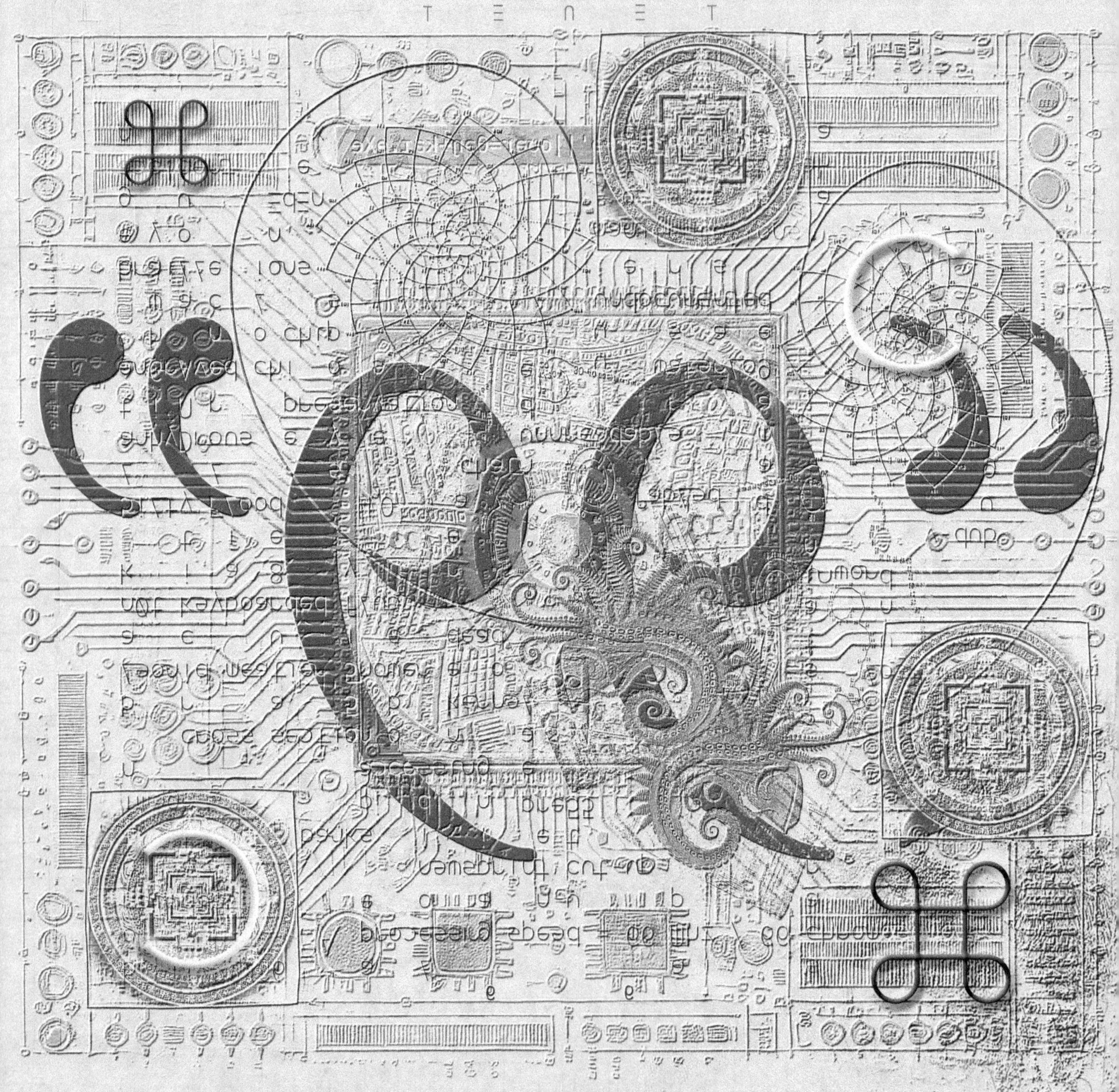

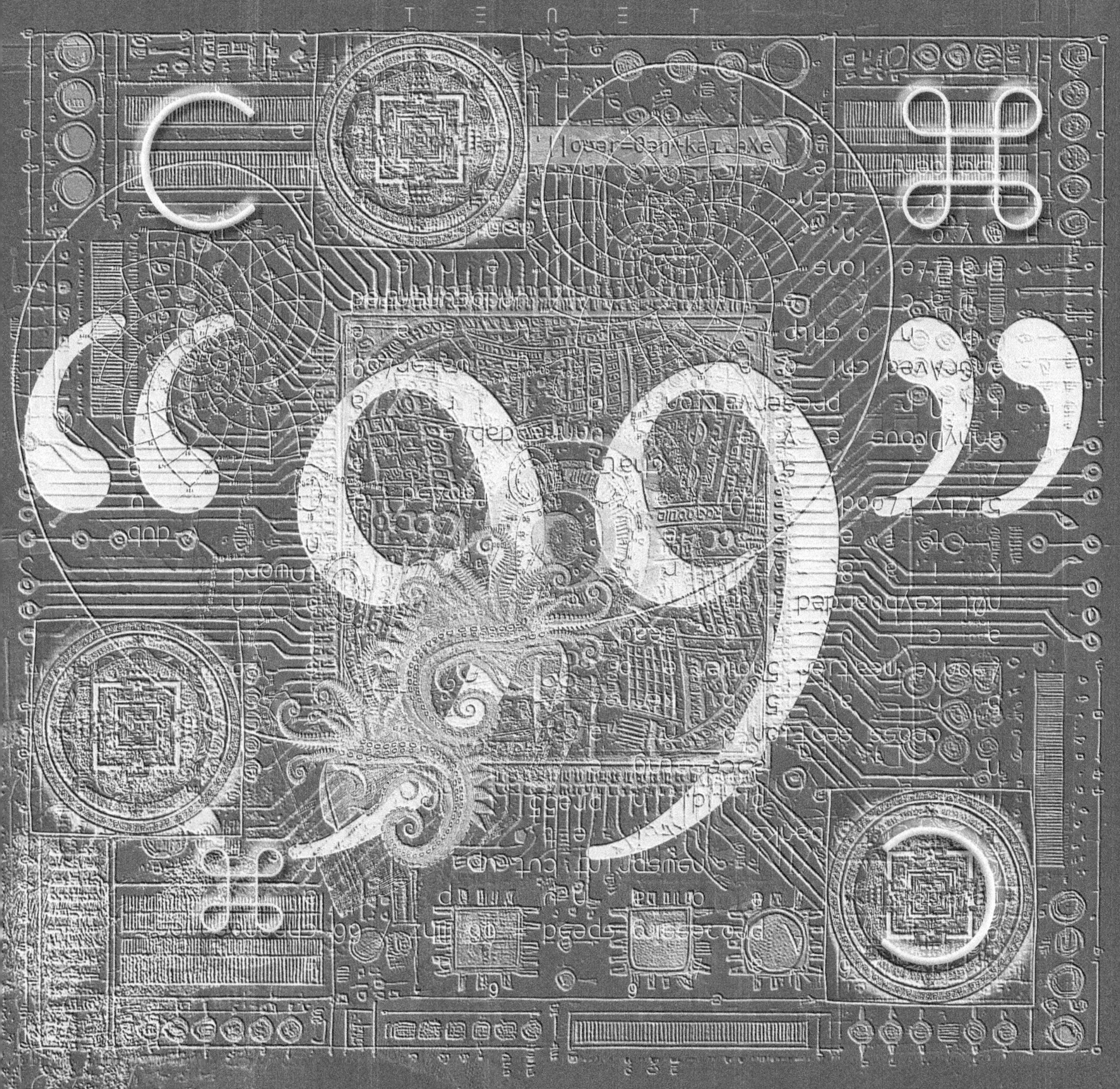

www.ingramcontent.com/pod-product-compliance
Lightning Source LLC
Chambersburg PA
CBHW082249060726
47592CB00022B/3338